International Virtual Conference on Recent Trends in Materials Science (IVCRTMS 2020)

IOP Conference Series: Materials Science and Engineering
Volume 1124

Karnataka, India
16 November 2020

Editors:

Augustine George
A. Sevuga Pandian
M. Ambrose Rajkumar
Anoop Baby K.B.

ISBN: 978-1-7138-3407-6
ISSN: 1757-8981

Printed from e-media with permission by:

Curran Associates, Inc.
57 Morehouse Lane
Red Hook, NY 12571

Some format issues inherent in the e-media version may also appear in this print version.

Institute of Physics
Dirac House, Temple Back
Bristol BS1 6BE UK

Phone: 44 1 17 929 7481
Fax: 44 1 17 920 0979

techtracking@iop.org

Additional copies of this publication are available from:

Curran Associates, Inc.
57 Morehouse Lane
Red Hook, NY 12571 USA
Phone: 845-758-0400
Fax: 845-758-2633
Email: curran@proceedings.com
Web: www.proceedings.com

TABLE OF CONTENTS

IOP Conf. Series: Materials Science and Engineering 1124 (2021) 011001 doi:10.1088/1757-899X/1124/1/011001

Proceedings of the International Virtual Conference on Recent Trends in Materials Science IVCRTMS 2020

November 16th, 2020

Organized by

DEPARTMENT OF PHYSICS
KRISTU JAYANTI COLLEGE (AUTONOMOUS)
K. Narayanapura, Bangalore – 560077
Karnataka, India

Editors

Rev. Dr. Augustine George
Prof. A. Sevuga Pandian
Dr. M. Ambrose Rajkumar
Dr. Anoop Baby K.B

IOP Conf. Series: Materials Science and Engineering 1124 (2021) 011001 doi:10.1088/1757-899X/1124/1/011001

Message From Principal

Materials make modern life more comfort—from the polymers in the chair you're sitting on, the metal ball-point pen you're using, and the concrete that made the building you live or work in, to the materials that make up streets and highways and the car you drive. All these items are products of materials science and technology. Good science produces theories that are explored through experimentation and the experiments depend upon the theories for direction. The discipline of Physics of materials is one such technology with ancient roots. The fundamental feature of human intelligence is creativity. New ideas can be created by producing novel combinations of familiar ideas, by exploring the potential of conceptual spaces and by making transformations that enable the generation of previously impossible ideas.

I congratulate the Department of Physics for the initiatives taken to bring the researchers across the nation in a common platform. I am confident that this conference will provide a stage to present and share the research findings for further enhancement and perfection. I am delighted to note that the accepted papers are being published in *IOP Conference Series: Materials Science and Engineering (MSE) – a* SCOPUS indexed journal.

I wholeheartedly appreciate all the sincere efforts of the entire team of IVCRTMS 2020 and wish them a grand success.

IOP Conf. Series: Materials Science and Engineering 1124 (2021) 011001 doi:10.1088/1757-899X/1124/1/011001

Message From Dean

Materials science combines many areas of science such as chemistry, physics, and engineering to make better, more useful, and more economical and efficient "stuff". Because of the interdisciplinary nature of materials science, it can be used both science and engineering.

IVCRTMS 2020 strives to be one of the most far-reaching meetings concentrated on the different aspects of Physics and Materials Science. It is certain to be the stage for scholastic and industry experts to examine ongoing advancement in the fields of physics, materials science and allied fields. The reason for this gathering is to make a logical commitment to the field of physics through dialogues and publications.

Special appreciation to the convenors and the organizing committee. We trust the delegates can acquire useful information from the proceedings. We look forward to the contributions during discussions and publications providing the foundation for future developments in this field.

On behalf of the Organizing and Advisory Committee we take great pleasure in welcoming the delegates to the conference to foster the progress in the field by contributing with your expertise.

IVCRTMS 2020 IOP Publishing
IOP Conf. Series: Materials Science and Engineering 1124 (2021) 011001 doi:10.1088/1757-899X/1124/1/011001

Message From Conference Chair

Technology has taken the forefront surpassing agricultural transformation and industrial revolution. The fields of Physics and Materials Science has expanded to include numerous parts of processing and innovation, and these terms are more conspicuous than any other time in recent memory.

The International Virtual Conference on Recent Trends in Materials Science (IVCRTMS 2020) conceives to unite interdisciplinary analysts researching the interface between Physics and Materials Science. Discussions as this will be instrumental in creating novel strategies for research and development. We sincerely believe this virtual gathering will be a platform for scientific exchange of ideas and to provide an opportunity for the dissemination of the latest results in research among the range of experts in the field Physics and Materials Science.

I sincerely hope that one day of healthy interactions and proactive exchange of ideas will prove to be fruitful and contribute immensely to our mutual growth. I congratulate the participants, Delegates, Organizing Committee, faculty members and students for their outstanding contribution in organizing this conference successfully.

Message From Conveners

This virtual conference aims at disseminating knowledge in the domains of Science and Engineering. The Conference will be an opportunity to reflect on what has been happening in the field of Science and Engineering, to share the experiences, perspectives and responses to the challenges faced by the researchers, and to give thought to set new directions for the future. The Conference will provide a platform for experts from academia and industry, to showcase their latest findings in the areas of expertise, that include physics and chemistry of advanced materials in Medical, Engineering, Nanotechnology, Environmental Technologies, Biomaterials and Green Chemistry. It is also a forum for the dissemination of original research findings, new ideas and practical experiments, which focus on both theory and practice.

The conference is expected to disseminate the global knowledge in the domain of Materials Science which covers the sub-topics of: Crystal growth and Crystallography / Nanoscience and Technology / Bio and Nanosensors / Solar cells and Photovoltaic devices / Energy Storage Materials / Polymers and Nanocomposites.

We extremely thank our management, Keynote speakers, advisory committee members, participants, reviewers, session chairs, organizing committee Members and all those who have helped us to organize this International Conference and to make it a success.

We thank our publication partner IOP Science Publishing for publishing the accepted papers in the *IOP Conference Series: Materials Science and Engineering (MSE) – a* SCOPUS indexed journal.

Dr. Ambrose Rajkumar M.
Dr. Anoop Baby K. B.

IOP Conf. Series: Materials Science and Engineering 1124 (2021) 011001 doi:10.1088/1757-899X/1124/1/011001

Conference Committee

CHIEF PATRON

Fr. Dr. Augustine George, CMI
Principal

ADVISORY COMMITTEE

Rev. Fr. Som Zacharias, CMI
Rev. Fr. Emmanuel P.J, CMI
Dr. Calistus Jude, Dean of Sciences

CONFERENCE CHAIR

Rev. Fr. Lijo P Thomas
Head,

Department of Computer Science (UG)

CONFERENCE CO-CHAIR

Prof. Sevuga Pandian
Staff Coordinator,
Department of Computer Science (UG)

CONFERENCE CONVENOR

Dr. M. Ambrose Rajkumar
Assistant Professor,
Department of Physics

Dr. Anoop Baby KB
Assistant Professor,
Department of Physics

Dr. Mayank Pandey
Assistant Professor,
Department of Physics

ADVISORY BOARD

Dr . D. Prem Anand, HOD of Physics,
St. Xavier's College (Autonomous),
Palayamkottai, Tamil Nadu.

Dr . M. Senthil Pandian, Research Scientist,
SSN College of Engineering,
Kalavakkam, Tamil Nadu.

Dr . R. Ramesh Babu, Associate Professor,
Bharathidasan University, Tiruchirappalli,
Tamil Nadu.

Dr . S. M. Ravi Kumar, Assistant Professor,
Sri Subramaniya Swamy Govt. Arts College,
Tirutani, Tamil Nadu.

Dr . S. Anbarasu, Assistant Professor,
St. Joseph's College (Autonomous),
Tiruchirappalli, Tamil Nadu.

Dr . G. Markandeyulu, Professor,
Indian Institute of Technology Madras,
Chennai, Tamil Nadu.

Dr . V. Srinivas Professor,
Indian Institute of Technology Madras,
Chennai, Tamil Nadu.

Dr . P. S. Anil Kumar, Professor,
Indian Institute of Science, Bangalore.

Dr . Satyam Suwas, Professor,
Indian Institute of Science, Bangalore.

Dr . S. Angappane, Scientist,
Centre for Nano and Soft Matter Sciences,
Bangalore.

Peer review declaration

All papers published in this volume of *IOP Conference Series: Materials Science and Engineering* have been peer reviewed through processes administered by the Editors. Reviews were conducted by expert referees to the professional and scientific standards expected of a proceedings journal published by IOP Publishing.

- **Type of peer review: Single-blind**
  The received research articles will be peer reviewed by expert researchers in the respective fields. Based on the comments and recommendation of the expert reviewer the paper will be accepted for publication
- **Conference submission management system:**
  Research articles were received through Google form which was given in the conference website
- **Number of submissions received** 23
- **Number of submissions sent for review** 23
- **Number of submissions accepted** 16
- **Acceptance Rate (Number of Submissions Accepted / Number of Submissions Received X 100) : 69.56 %**
- **Average number of reviews per paper** 01
- **Total number of reviewers involved** 07
- **Any additional info on review process :**

Authors were received comments and suggestions from the reviewer for each research articles separately. All received papers were reviewed by the experts in the respective area of research. Authors were received comments related to technical, experimental, grammatical, and analytical aspects of their research. The required modifications were suggested by the reviewers. Reviewers have recommended the manuscripts for publication after incorporation of the suggested changes. Authors were provided with the period of seven days to do the modifications as per suggestions given by the reviewers. Modified papers were sent to the reviewer for verification and papers were accepted as per the report given by the reviewers.

Contact person for queries

: Dr. M.Ambrose Rajkumar
Assistant Professor of Physics
Kristu Jayanti College (Autonomous)
K. Narayanapura, Kothanur Post,
Bengaluru-77, Karnataka
Mobile: +91- 89733 14608_
ambrose@kristujayanti.com

Dr. Anoop Baby. K.B
Assistant Professor of Physics
Kristu Jayanti College (Autonomous)

IOP Conf. Series: Materials Science and Engineering 1124 (2021) 011002 doi:10.1088/1757-899X/1124/1/011002

K. Narayanapura, Kothanur Post,
Bengaluru-77, Karnataka
Mobile: +91- 99626 04867_
anoopbaby@kristujayanti.com

IOP Conf. Series: Materials Science and Engineering 1124 (2021) 012001 doi:10.1088/1757-899X/1124/1/012001

Study of adsorption of methylene blue in aqueous solution onto activated carbon prepared using low cost domestic oven based heating system

Sinmay Sahoo[1], Maya Devi[1] and Susanta Kumar Das[2*]
[1] Dept. of Physics, School of Applied Sciences, Kalinga Institute of Industrial Technology, Deemed to be University, Bhubaneswar-751024, Odisha, India
[2] School of Physics, Gangadhar Meher University, Sambalpur-768001, Odisha, India

Abstract. In this work activated carbon was prepared from teak wood charcoal using very low-cost oven based heating system. Firstly, charcoals were grinded with a domestic grinder and then activated with 25% $CaCl_2$ solution and then heated with 250C for different times. And then experiment was carried out through 2.5×10^{-5} methylene blue aqueous solution by taking activated Charcoal into it and kept one hour for adsorption. Adsorption capacity was estimated by taking the different amount of activated carbon. The result indicates that prepared activated carbon be effectively used for water purification application through adsorption of the dye.

1. Introduction

Activated carbon is a form of carbon that has been processed with oxygen to create millions of tiny pores between the carbon atoms. The internal surface area of commercial activated carbon is ranging from 500 to 1500 mm^2/g. Because of its large surface area and high adsorption capacity it has been used for various applications like medical uses, adsorption of natural gas, cosmetics use, for removal of heavy metals, water purification etc. As per the demand of human life, industrial revolution polluting our lakes, oceans and rivers. For which it is damaging our planet and organism die very rapidly. So, water Purification is major and is the most essential component for our life. The deficiency to access clean and fresh water can lead many troubles including sickness, malnutrition and death. So, it is becoming most important for future due to the increase in population, changes in climate, expansion of agricultural and industrial activities with enhancing the living standards. The contamination of water occurs because the industries release many toxic organic dyes, heavy metals and micropollutants into water.

Methylene blue used as a traditional dye for silk wool and cotton also used to make certain body fluids and tissues easier to view during surgery, is one of the toxic dye. Consuming water which contains methylene blue may cause vomiting, mild bladder irritation, dizziness, diarrhoea and the burn effect of eyes. Recently different techniques have already been used for the removal of dye from water and wastewater are biodegradation [1], electrochemical oxidation [2], solvent extraction [3], ion exchange biological purification [4], coagulation & sedimentation techniques and adsorption [5]. However, most of these techniques are comparatively expensive and complicated. But adsorption technique has been preferred because it is quite easy and cost effective as adsorbents, porous solids which generally have a high surface area. So now a days it is quite challenging for developing countries to produce high-efficient and low cost activated carbon. But many problems arise at the time of regeneration of used activated carbon. Many commercial activated carbon has been widely used for the adsorption process.

[1] sinmay1990sahoo@gmail.com

However, from economical point of view, it has expensive in production process [6]. So, producing low-cost and effective activated carbon may give a healthy environmental to our society and offer benefits for future commercial applications. The activated carbon prepared from unused biomaterials is very low cost compared to the cost of commercial activated carbon. Many people obtained activated carbons from agricultural by-products including soybean straw [7], coconut shell [8], date pit [9], rattan sawdust [10], and sewage sludge [11]. But to produce activated carbon by using very low cost equipment's is quite interesting topic for us. Various works have been done to produce low cost activated carbon with good approach. But they don't look into the matter of low cost oven based heating system. A. Ghanadzadeh Gilani *et al* prepared the activated carbon for phenol adsorption using sealed ceramic oven [12]. Tchuifon Tchuifon D. Raoul *et al* used an ISUNU mark electric furnace with automatic regulation to prepare the activated carbon for phenol adsorption [13]. Md. Masuduzzaman *et al* used infrared cooker to prepare activated carbon for methylene blue adsorption [14]. Sadaf Adhami *et al* used tubular reactor as the heating system to prepare activated carbon for water treatment [15]. Wankhade Amey A. *et al* used muffle furnace as the heating system for the carbonisation of activated carbon [16]. However, all these heating systems are little bit expensive. The main aim of this study is to use low cost equipment's for the preparation of activated carbon and use it for water purification application through adsorption of the dye.

2. Experimental

2.1. Preparation of Activated Carbon

The activated carbon was prepared from the teak wood charcoals by following the standard procedure. This procedure consists of 5 steps.

Step 1- Collection of charcoals:

The teak charcoals (Figure 1(a)) were purchased from a local vendor.

Step 2- Grind up the charcoals:

The charcoals were cleaned by water. 250 gm of cleaned charcoal were grinded by using a domestic mixture grinder. The charcoal powder is left for 24 hours to air dried completely.

Step 3- Preparation of solution for activating the charcoal powder :

For activation of the charcoal, a solution was prepared by mixing calcium chloride and water in 1:3 ratio. For ours case 50 gm of calcium chloride was mixed with 150 ml of water.

Step 4- Preparation of charcoal paste :

A slurry type of paste was prepared. First 50 gm of dry charcoal powder was taken in a beaker. 50 ml of calcium chloride solution was added to it and stirring was done with a spoon until getting slurry paste. Then this paste was kept for 24 hours in dark.

Step 5- Activation of the sample by the low cost oven based heating system :

After 24 hours these paste was first dried in the sunlight for 3 hours, then rinsed with clean water. Then it was heated at 250 degrees Celsius for certain time. Three set of samples S1, S2 and S3 were prepared with heating time of 30 min, 60 min and 90 min respectively. A domestic oven-based heating system was used for this propose. Figure 1(b) was the prepared activated carbon.

IOP Conf. Series: Materials Science and Engineering 1124 (2021) 012001 doi:10.1088/1757-899X/1124/1/012001

1(a)

1(b)

Figure 1(a). Charcoal used for preparation of activated carbon.
Figure 1(b). Prepared activated carbon.

2.2. Water purification application

Water purification experiments were done by exploiting the adsorption properties of activated carbon. For this experiment, every time 50 ml of aqueous solution of methylene blue (MB) (molar concentration = 2.5 x 10^{-5} M) and some specific amoumt of activated carbon were taken in a beaker. This mixture was then stirred gently for adsorption. This process continued till getting the appropriate amount of activated carbon for complete adsorption of methylene blue dye. The adsorption capacity of methylene blue q (mg of adsorbate/g of adsorbent) was calculated by using the following equation

$$q=\frac{(C_0 - C_m)V}{m} \qquad (1)$$

where V is the volume of the solution in terms of L, C_0 the initial concentration (mg l^{-1}), C_m the final concentration (mg l^{-1}), and m is the weight of the adsorbent (g).

$$\text{And \% of removal of MB} = \frac{(C0 - C)}{C0} \times 100 \qquad (2)$$

Where C_0 is the initial concentration and C is the concentration of the aqueous solution at the time of measurement.
The absorbance spectroscopy of the dye solutions were done to find the ratio of the dye concentration.

3. Results and Discussion

3.1 Study of removal of MB by using different amount of activated carbon prepared by heating for 30 min.

The Figure 2 shows the percentage of the dye removal with different amount of activated carbon prepared by heating for 30 min. From this it is found that when we added more amount of adsorbent (activated carbon) more percentage of removal of methylene blue dye has taken place. This figure also shows that for complete removal of dye the minimum required amount of activated carbon is 0.06g. We got a flat curve after this weight. This indicates that further addition of the activated carbon is not required. The color of the dye solution without adsorption of the dye (for the case of no activated carbon) and complete adsorption of dye are shown in Figure 3(a) and 3 (b) respectively. It is to be informed here that without activation no significant amount of dye adsorption has been found in the similar experimental condition.

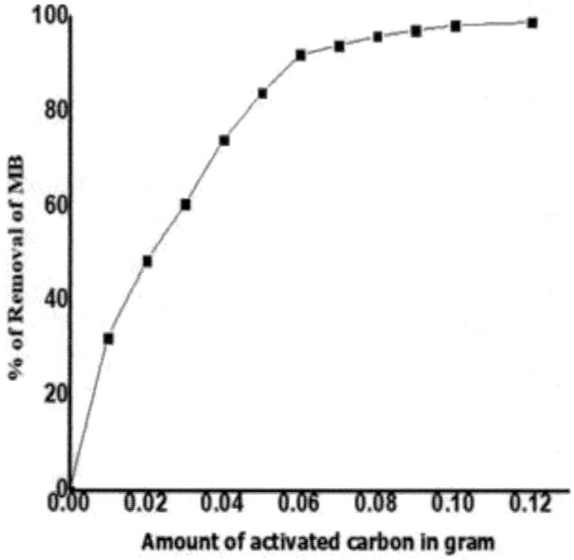

Figure2. Percentage of the dye removal with different amount of activated carbon.

3(a)

3(b)

Figure 3(a). Color of the dye solution without adsorption of the dye (for the case of no activated carbon).
Figure 3(b). Complete adsorption of dye (for the case of no activated carbon ≥ 0.06 g).

3.2 Comparison of adsorption capacity of activated carbon prepared by heating for 30 min, 60 min., 90 min.

Considering the data of incremental region of % of dye removal (Please ref. data points corresponding to activated carbon of weight within the range of 0-0.06 gm of Figure 2), the adsorption capacity for the activated carbon prepared by heating for 30 min is found to be 6.66 mg/g. In the same manner the adsoption capacity for the case for 60 min and 90 min were found to be 6.15 mg/g and 2.85 mg/g respectively. In other word we can say, the adsorption capacity for 60 min of heating case was almost found to be same as the case of 30 min heating. But for 90 min of heating process it is found to be less than half of that for the case of 30 min heating. The most probable cause for this is probably because of reduction of surface area due to coagulation. Further investigation is going on this regards.

.

4. Conclusion

In this work we have demonstrated the complete removal of the methylene blue dye with the activated carbon prepared by domestic oven based heating system. The activated carbon produced at 250C activation temperature for 30 mins shows the best adsorption behaviour. The cost of the domestic oven used for activation was only about $20. So, in this work demonstrate a cost effective ways of activation.

5. References

[1]. El-Naas M H, Al-Muhtaseb S A, Makhlouf S, 2009 Biodegradation of phenol by Pseudomonas putida immobilized in polyvinyl alcohol (PVA) gel *J. Hazard. Mater.* **164** 720-725

[2] Yavuz Y, Koparal A S, 2006 Electrochemical oxidation of phenol in a parallel plate reactor using ruthenium mixed metal oxide electrode *J. Hazard. Mater.* **136** 296-302

[3] Yang C, Qian Y, Zhang L, Feng J, 2006 Solvent extraction process development and on-site trial-plant for phenol removal from industrial coal-gasification wastewater *Chem. Eng. J.* **117** 179-185

[4] Caetano M, Valderrama C, Farran A, Cortina J L, 2009 Phenol removal from aqueous solution by adsorption and ion exchange mechanisms onto polymeric resins *J. Colloid. Interf. Sci.* **338** 402-409

[5] Yousef R I, El-Eswed B, Ala'a H, 2011 Adsorption characteristics of natural zeolites as solid adsorbents for phenol removal from aqueous solutions: kinetics, mechanism, and thermodynamics studies *Chem. Eng. J.* **171** 1143-1149

[6] Özkaya B, 2006 Adsorption and desorption of phenol on activated carbon and a comparison of isotherm models *J. Hazard. Mater.* **129** 158-163

[7] Mohammad Y S, Egbenya M, Sunday B I, Abdul-Raheem G, Charles A O, 2015 Isotherm, kinetics and thermodynamics of phenol adsorption onto rice husk activated carbon *Leonardo El. J.* Pract. *Technol*. **14** 115-128

[8] Din A T M, Hameed B, Ahmad A L, 2009 Batch adsorption of phenol onto physiochemical-activated coconut shell *J. Hazard. Mater.* **161** 1522- 1529

[9] Tseng R L, Wu F C, juang R S, 2003 Liquid-phase adsorption of dyes and phenols using pinewood-based activated carbons *Carbon* **41** 487-495

[10] Hameed B, Rahman A, 2008 Removal of phenol from aqueous solutions by adsorption onto activated carbon prepared from biomass material *J. Hazard. Mater.* **160** 576-581

[11] Li D, Wu Y, Feng L, Zhang L, 2012 Surface properties of SAC and its adsorption mechanisms for phenol and nitrobenzene *Bioresource Technol.* **113** 121-126

[12] Ghanadzadeh Gilani A, Ghanadzadeh Gilani H, Azmoon P, Chaibakhsh N, 2019 Phenol Adsorption from Aqueous Phase onto Prepared Low-cost Carbons from Natural Sources: A Comparative Study *Phys. Chem. Res.* **7** 327-346

[13] Tchuifon Tchuifon D R, Anagho S G, Ketcha J M, Nche G Ndifor-Angwafor And Ndi J N, 2014 Kinetics and equilibrium studies of adsorption of phenol in aqueous solution onto activated carbon prepared from rice and coffee husks *Int. J. Eng. Tech. Research* **2** ISSN: 2321-0869

[14] Md. Masuduzzaman, Rajesh Kumar Ghosh, Shuvagata Biswas, Dr. Bodius Salam, 2017 Production of Activated Carbon from charcoal using chemical activation *International Conference on Mechanical Engineering and Renewable Energy* 2017 (ICMERE 2017-PI-142)

[15] Sadaf adhami , Hamed Ghorbanpoor, Berat azak , Serkan kapucu, Macid Nurbas, Hüseyin AVCI 2018 A Novel Approach For Water Treatment by using Activated Carbon: Apricot Kernel

Shell, *The Journal of Engineering and Architecture Faculty of Eskisehir Osmangazi University* **26(1)** 1-7

[16] Wankhade Amey A, Ganvir V N, 2003 Preparation of Low Cost Activated Carbon from Tea Waste using Sulphuric Acid as Activating Agent *Int. Res. J. Env. Sci.* **2(4),** 53-55

[17] Stasinakis A S, Elia I, Petalas A V, Halvadakis C P, 2008 Removal of total phenols from olive-mill wastewater using an agricultural by-product, olive pomace *J. Hazard. Mater.* **160** 408-413

[18] Yahia A A, 2006 Activated carbon from dates stone by $ZnCl_2$ activation *Eng. Sci.* **17(2)** 75-100

[19] Yuan C S, Lin H Y, Wu C H, Liu M H, 2004 Preparation of sulfurized powdered activated carbon from waste tires using and innovative compositive impregnation process *J. Air & waste Manage. Assoc.* **54** 862-870

[20] Dina D J D, Ntieche A R, Ndi J N, Ketcha M J, 2012 Adsorption of acetic acid onto activated carbons obtained from maize cobs by chemical activation with zinc chloride ($ZnCl_2$) *Res. J. Chem. Sci.* **2 (9)** 42-49

[21] Baçaoui A, Yaacoubi A, Dahbi A, Bennouna C, Luu R P T, Maldonado-Hodar F J, ... & Moreno-Castilla C, 2001 Optimization of conditions for the preparation of activated carbons from olive-waste cakes *Carbon* **39(3)** 425-432

[22] Balci S, Doğu T, Yücel H, 1994 Characterization of activated carbon produced from almond shell and hazelnut shell *J. chem. Tech. biotech.* **60(4)** 419-426

[23] Bevla F R, Rico D P, and Gomis A F M, 1984 Activated Carbon from Almond Shells. Chemical Activation. Activating Reagent Selection and Variables Influence *Ind. Eng. Chem. Prod. Res. Dev.* **23** pp.266-269

[24] Alhamed Y A, 2009 Adsorption kinetics and performance of packed bed adsorber for phenol removal using activated carbon from dates' stones. *J. Hazard. Mater.* **170** 763-770

[25] El Hannafi N, Boumakhla M, Berrama T, Bendjama Z, 2008 Elimination of phenol by adsorption on activated carbon prepared from the peach cores: modelling and optimisation *Desalination* **223**, 264-268

[26] Basso M C, Cerrella E G, Cukierman A L, 2002 Activated carbons developed from a rapidly renewable biosource for removal of cadmium (II) and nickel (II) ions from dilute aqueous solutions *Industrial & Eng. Chem. research* **41(2)** 180-189

[27] Subhashree P, 2011 Production and characterization of activated carbon producted from a suitable industrial sludge *Projet report, National Institute of Technology Rourkela*, pp 32-33

[28] Lü G, Hao J, Liu L, Ma H, Fang Q, 2011 The adsorption of phenol by lignite activated carbon *Chinese. J. Chem. Eng.* **19** 380-385

[29] Ho Y S, Porter J F, McKay G, 2002 Equilibrium isotherm studies for the adsorption of divalent metals ions onto peat: copper, nickel and lead single component systems *Water, Air and Soil pollut.* **141** 1-33

[30] Srihari V, Das A, 2008 The kinetic and thermodynamic studies of phenol-sorption onto three agro-based carbons *Desalination* **255** 220-234

[31] Hall K R, Eagleton L C, Acrivos A, Vermeulen T, 1996 Pore and solid-diffusion kinetics in fixed-bed adsorption under constant-pattern conditions *Ind.Eng.Chem.Fund.* **5** 212-223

IOP Conf. Series: Materials Science and Engineering 1124 (2021) 012002 doi:10.1088/1757-899X/1124/1/012002

Polaronic Relaxation and Variable-Range-Hopping Conductivity in Fe_3O_4 Nanoparticles

A K Swetha, B H Pradeep, R Mallikarjun, S Singh and R S Joshi[1]

Central University of Karnataka, Kalaburagi-585368, India

Abstract. We report the polaronic relaxation and Variable Range Hopping (VRH) conductivity in 100nm average particle size cold pressed Fe_3O_4 nanoparticles. The crystal structure was studied using XRD showing spinel structure with (311) intense peak. The AC transport in the cold pressed powder was studied from 10 K to room temperature and from 100 Hz to 5.5 MHz. It was observed that the solid couples capacitively with the perturbing electric field initially and as the frequency increases it becomes more inductive. The impedance data was modelled using modified Havriliak–Negami relaxation model. It was observed that the cold pressing introduces an asymmetric broadening in the relaxation process. The change in impedance with temperature showed a drastic drop at around 120 K due to Verwey transition in real as well as imaginary part. The change in DC resistance with temperature indicated VRH mechanism of charge transport in the solid with a metal insulator transition at around 120 K. The relaxation times calculated indicated that the hopping process has an attempt frequency of the order of 30 microseconds. The activation energy change across the phase transition with relaxation time and resistance indicated that the charge transport is coherent. The activation energy of ~0.05eV above Varway transition indicated that the transport is small polaronic in nature.

1. Introduction

Half metallic solids are extensively studied for many years since these are an ideal component for spintronic devices. These show spin dependent scattering and strong electron-electron correlation, resulting in metal insulator phase transition [1]. Bulk Fe_3O_4 exhibits half metallicity with theoretically 100% spin polarization [2]. The dynamic magnetic and semiconducting properties provide a prominent role to Fe_3O_4 in the development of oxide electronics [3]. Fe_3O_4 has mixed iron valences of Fe^{+3} and Fe^{+2}. It shows inverse-spinel cubic structure(AB_2O_4) at ambient conditions. Upon cooling below a certain temperature it shows exceptional discontinuous drop in the electrical conductivity termed as Verwey Transition (T_V). The metallic nature at T> T_V could be explained by the charge disorder (CD) within the B sites as well as the metal-insulator transition below T_V is due to the charge ordering (CO) [4]. The electrical transport in Fe_3O_4 is drastically affected by the dimensionality of the solid. At lower dimensions the Tv as well as the electrical resistance gets affected by the particle size in nanoparticles [5] as well as thickness of the films [6]. The measurement of electrical transport at lower dimensions preserving the dimensionality is a challenge. The heat treatment to the powder or film for compacting changes the very governing parameters like particle size or grain size leading to loss of dimensionality effect [7].

[1]rajeevsj@cuk.ac.in

In the present work we investigate the charge transport properties of Fe_3O_4 nanoparticles in cold pressed condition preserving the effect of the dimensionality [8], which is difficult to achieve if the particles are thermally treated. The Fe_3O_4 powder of average 100nm size is subjected to dielectric relaxation study from 10 K to room temperature. The relaxation was modelled using asymmetric-dissipative Havriliak–Negami (HN) dispersion model [9]. The relaxation time and the asymmetry parameters are studied to interpret the effect of charge ordering across T_v with variation of frequency.

2. Experimental

The 99 % pure Fe_3O_4 nanoparticles having average size 100 nm were commercially purchased from Sigma Aldrich. The crystal structure of the same was studied by the PROTO AXRD Powder diffractometer. The patterns with Cu Kα radiation (λ=1.5405 Å) were recorded in the region 20° to 80°. The powder was cold pressed and compacted for 30 minutes under pressure. The pellet formed was coated with silver to form electrodes. The low temperature electrical measurements were done in Advanced Research Systems, Inc. CCR (Closed Cryogenic Refrigerator), along with lakeshore temperature controller model 335. To analyse the impedance response, ZM2376 LCR meter of frequency range from 1mHz to 5.5 MHz was used. Parameters like R_P, X, C_P, D were measured on various temperatures from room temperature to 10 K. To avoid electrode effects the measurements are accounted only from 100 Hz to 5.5 MHz.

3. Results and Discussion

Figure 1 shows the XRD pattern of Fe_3O_4 nanoparticles with high intense peaks at (311) crystal plane with no characteristic peaks of impurities (indexed with JCPDS 19-0629) which infers the inverse cubic spinel structure. The particle size was found using the Debye-Scherer formula $d = K\lambda/(\beta\cos\theta)$ as 28.85nm [10,11]. The grade of crystallinity in these nanoparticles was found higher with minimal hump in the background. This information about the average crystallite size is expected to affect the electrical transport and dielectric relaxation significantly [12] as evident from further measurements.

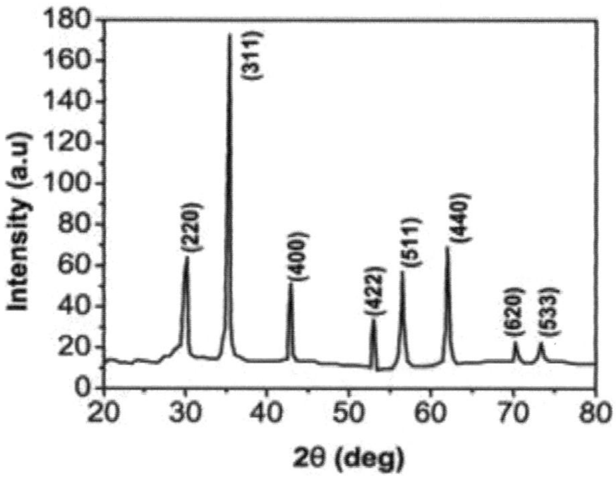

Figure1. The XRD pattern of Fe_3O_4 nanoparticles.

The variation of real part of impedance (R) with frequency was obtained at various temperatures for Fe_3O_4 nanoparticles. From figure 2 it is clear that the magnitude of R decreases when temperature increases which corresponds to a thermally activated type of conductivity [13]. At low temperature R increases sharply indicating an insulating behaviour. With frequency the resistance decreases indicating a pronounced capacitive coupling. As the frequency increases the resistance decreases inferring increased inductive coupling. The relaxation or change in resistance was found to be broad which is a typical feature of semiconducting nanoparticles [14]. With increase of temperature the metallicity in the particles was found to increase.

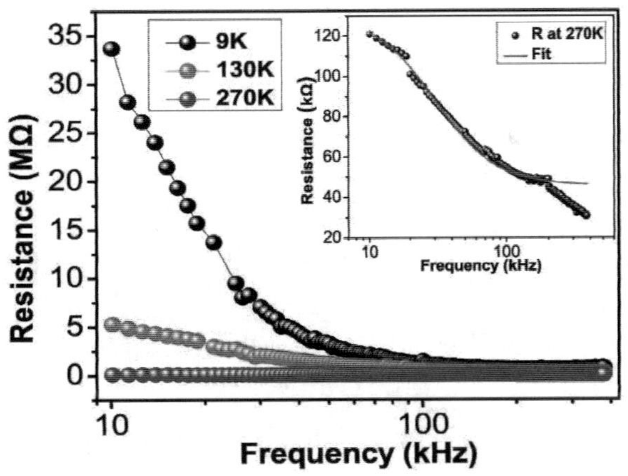

Figure 2. The variation of real part of impedance with frequency for various temperatures. The inset image shows fit to the modified HN Model.

The analysis of the impedance relaxation processes was done for the inhomogeneous medium with modified HN method of relaxation applied to real part of the impedance (Since R is also a valid response function). It is important to note that the cold pressed powder can be treated as an inhomogeneous medium with discontinuities and defects. Since the HN model was used to represent both symmetric and asymmetric broadening with variation in the values of α and β it is a valid choice for this system. It can be expressed for impedance of a medium as a function of frequency,

$$R(\omega) = R_\infty + \frac{\Delta R}{\left(1 + (i\omega\tau)^\alpha\right)^\beta}$$

where, R_∞ represents the resistance at the high frequency limit , $\Delta R = R_s - R_\infty$, R_s-static resistance and τ is the characteristic relaxation time.

Relaxation time τ is calculated from modified HN fit. τ value was found to be higher at lower temperature i.e., the frequency of relaxation is smaller, which means the electron cluster is tighter at these temperatures inferring the formation of charge ordered areas in the sample. It was realized that the broadening of the AC resistance with frequency increases indicating the loss of cluster area with increase in temperature. A sharp jump in the relaxation was observed at 30 K as shown in figure 3 indicating possibility of a blocking process in the partial volume of the nanoparticles [15]. It is interesting to note that the bulk resistance of the solid does not pick-up the feature of blocking whereas relaxation time does. It is essential to observe a magnetic transition being picked by an electrical transport measurement as indicated in this case. This could be equally a possibility of magnetoelectric coupling in the solid across the blocking temperature. It is known in literature that the magnetoelectric coupling can be seen in Fe_3O_4 like ferromagnetic solids across the phase transition which may involve a structural change as well as magnetic order change like spin glass [16]. The slow relaxation processes involving clustering and de-clustering of spins in low dimensional solids also can be investigated by this method [2].

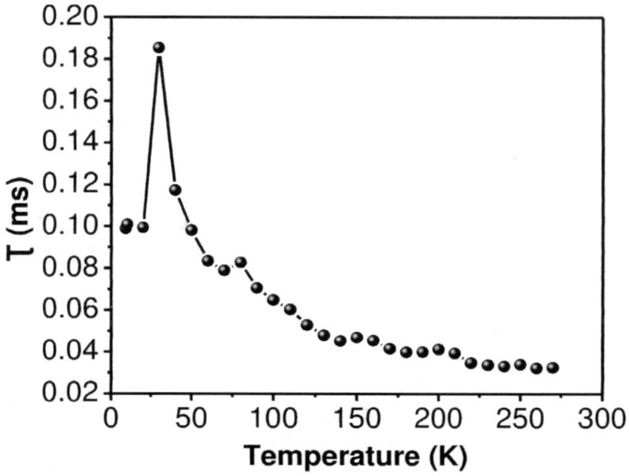

Figure 3. Relaxation time (τ) obtained from HN fit of the resistance plotted against temperature

τ is maximum at the lowest temperature above blocking, it decreases gradually when temperature is increasing and attains minimum at room temperature. This indicates melting of charge order and increased electron ion scattering with temperature in the solid. The drastic change in slope of the relaxation time with temperature [16] is attributed to insulator metal transition in the solid.

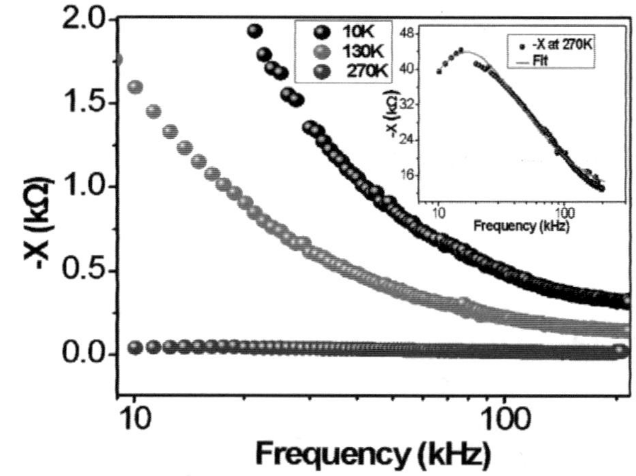

Figure 4. The variation of imaginary part of impedance with frequency for various temperatures. The inset image shows fit to the modified HN Model.

To pick up the dissipative processes in the solid the reactance or imaginary part of the impedance (X) is measured for any inhomogeneous medium. The variation of X with frequency at several chosen temperatures is shown in the figure 4. Wideness of the reactance peak indicates distribution of relaxation times in this material [13]. The decrease of reactance magnitude when temperature increases, suggests a thermally generated carrier contribution to electrical conductivity. Here, reactance decreases gradually with frequency. This is a clear indication of the capacitive response of the sample which depends on both temperature as well as nature of the material (internal electrical environment). Since τ is the time between two collisions, at high temperature, a kind of ballistic transport occurs as electrons are moving with very long distance collisions. It's a typical feature of any low dimensional system. Furthermore, one can notice a kink in relaxation time at about 30K obtained from reactance data supporting the possibility of blocking behaviour at low temperature in a selected part of the sample volume. It is interesting to observe a discontinuity in relaxation time across the T_v. The fact that reactance can pick up the dissipative processes better compared to the bulk resistance is evident in this observation shown in figure 5.

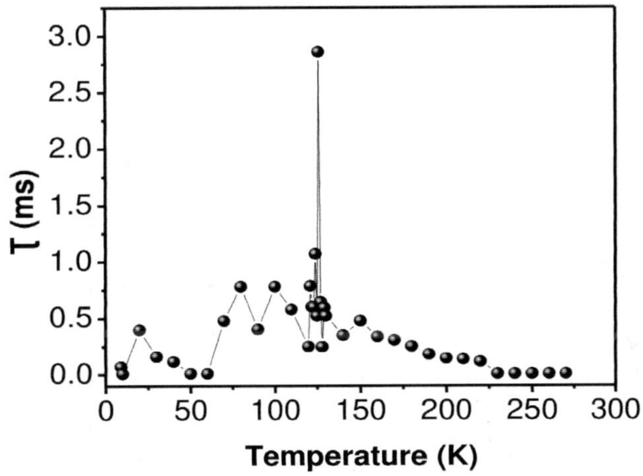

Figure 5. Relaxation time (τ) obtained from reactance with respect to temperature

The slope change is seen across the T_v with a peak-like behaviour corresponding to a significant electric coupling to the other properties like structure or magnetization.

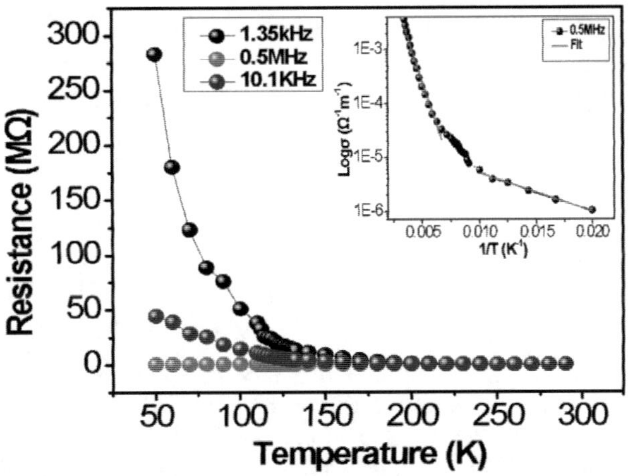

Figure 6. Variation of resistance with respect to temperature plotted for three different frequencies and the inset image is the variation of conductivity with temperature fitted with an Arrhenius model.

The temperature dependence of electrical resistance with different frequencies is studied from 10-273 K. The decrease in resistance with the increase in temperature indicates the insulator-metal transition (Verwey transition) and the transition is pronounced in the range of 120 K. We have found the activation energy (Ea) using the Arrhenius equation, by fitting the log ρ versus 1/T i.e.

$$\rho = \rho_0 exp\,(\frac{E_a}{KT})$$

where k, is the Boltzmann constant. Here the relaxation has the thermal activation with a barrier height of E_a[17]. The activation energy increases with increasing frequency as shown in figure 6. The fit for the selected frequency is shown in the figure. For this, Ea above Verwey was 56.79meV and 6.17meV below T_v. For bulk Fe_3O_4 activation energy is supposed to be close to 62.2 meV [18] above T_v which supports our observation. The increase of activation energy with frequency shows the coupling of electromagnetic radiation with the charge clusters in the inhomogeneous medium resulting in higher activation energy or local increase of barrier height above T_v.

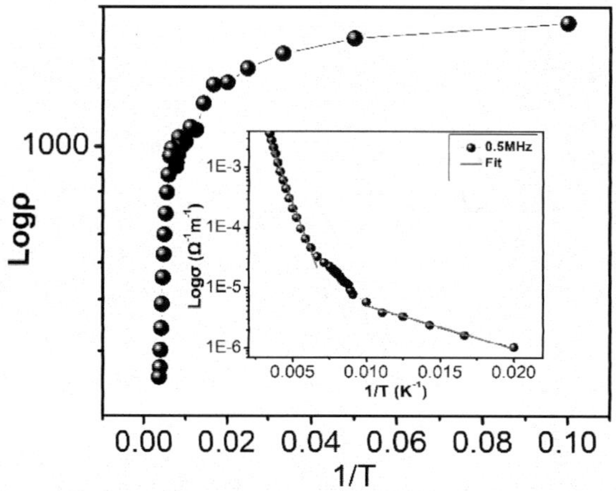

Figure 7. Variation of DC resistivity with temperature.

The DC electrical transport behaviour of bulk Fe_3O_4 was investigated in the temperature range of 10-273 K. log(ρ) vs 1/T curves plotted and the results were fitted to the general expression of activated transport to understand whether the conduction regime corresponding to a semiconductor or to a disordered metal [6].

$$\rho = \rho_0 \, exp(\frac{T_m}{T})^n$$

where ρ_0 is a constant that depends on the phonon density, T_m corresponds to Mott characteristic temperature related to the hopping barrier of the localized states. The exponent n decides the dimensionality (D) of transport in the system using the relation $= \frac{1}{1+D}$, the values of n are 1/2, 1/3 and 1/4 for 1D, 2D and 3D respectively[19]. In the present work the transport is divided into two regimes. The first with power n=0.5 in the temperature range 170 to 270 K where the dimensionality of transport is 1D the collective charge transfer being Shklovskii and Efros VRH [19]. Further in the low temperature range from 10 K to 120 K the dimensionality was not well defined because of clustered regions of charge order imposed with physical inhomogeneity in the solid.

4. Conclusion

We establish Shklovskii and Efros VRH kind of transport in the cold pressed inhomogeneous Fe_3O_4 above Tv. The relaxation time was found to be higher at lower temperature indicating formation of charge clusters in the insulating phase. The increase of activation energy with frequency indicates coupling electromagnetic radiation with the charge cluster producing higher electron scattering. The reduction of impedance with temperature infers the melting of charge order at higher temperatures. The relaxation time change with temperature supports the observation of charge order melting. The study provides insight into the charge transport mechanism in the physically inhomogeneous medium with magnetic, electrical as well as structural phase transitions specifically in Fe_3O_4 which is very important for spintronics.

5. References

[1].C.M. Fang, G.A. de Wijs, R.A. de Groot 2002 *Appl. Phys.* **91**, 8340-4 202
[2]. Coey J M D, Berkowitz A E, Balcells L I, Putris F F and Barry A 1998 Magnetoresistance of chromium dioxide powder compacts *Phys. Rev. Lett.* **80** 3815–8
[3].Kohiki S, Nakamura Y, Akiyama-Hasegawa K, Mitome M and Tsuya D 2018 Field Effect of a Chemically Assembled Fe_3O_4 Nanocrystal Film Single-Electron Transistor *Phys. Status Solidi Appl. Mater. Sci.* 215 1–7
[4].Markovich G, Fried T, Poddar P, Sharoni A, Katz D, Wizansky T and Millo O 2002 Observation of the Verwey transition in Fe_3O_4 nanocrystals *Mater. Res. Soc. Symp. - Proc.* 746 151–6
[5]Bae C J, Angappane S, Lee Y, Lee J, An K and Hyeon T 2007 in monodisperse Fe_3O_4 nanoparticles 4–6
[6].Bohra M, Roy Chowdhury D, Bobo J F and Singh V 2019 Anomalous electric transport across Verwey transition in nanocrystalline Fe_3O_4 thin films *J. Appl. Phys.* 125
[7]. Manheller M, Karthäuser S, Waser R, Blech K and Simon U 2012 Electrical transport through single nanoparticles and nanoparticle arrays *J. Phys. Chem. C* **116** 20657–65
[8]. Lee J, Kwon S G, Park J G and Hyeon T 2015 Size Dependence of Metal-Insulator Transition in Stoichiometric Fe_3O_4 Nanocrystals *Nano Lett.* **15** 4337–42
[9]Jonscher A K 1983 Dielectric relaxation in solids, *Chelsea Dielectrics Press, London*
[10]. Wei Y, Han B, Hu X, Lin Y, Wang X and Deng X 2012 Synthesis of Fe_3O_4 nanoparticles and their magnetic properties *Procedia Eng.* **27** 632–7
[11].Yu B Y and Kwak S Y 2010 Assembly of magnetite nanocrystals into spherical mesoporous aggregates with a 3-D wormhole-like pore structure *J. Mater. Chem.* **20** 8320–8
[12]. Köseoğlu Y and Kavas H 2008 Size and surface effects on magnetic properties of Fe_3O_4 nanoparticles *J. Nanosci. Nanotechnol.* **8** 584–90
[13]. Mahato D K, Molak A and Pawlus S 2020 Impedance, dielectric, and magnetic properties study of La_2CrMnO_6 ceramics *Ceram. Int.* **46** 6368–76
[14]. Sandler S E, Fellows B and Thompson Mefford O 2019 Best Practices for Characterization of Magnetic Nanoparticles for Biomedical Applications *Anal. Chem.* **91** 14159–69
[15]. Chen H, Billington D, Riordan E, Blomgren J, Giblin S R, Johansson C and Majetich S A 2020 Tuning the dynamics in Fe_3O_4 nanoparticles for hyperthermia optimization *Appl. Phys. Lett.* **117** 073702-4
[16].Manna K, Joshi R S, Elizabeth S and Anil Kumar P S 2014 Evaluation of the intrinsic magneto-dielectric coupling in $LaMn_{0.5}Co_{0.5}O_3$ single crystals *Appl. Phys. Lett.* **104** 2–6
[17]. G. A. Samara 1968 Effect of pressure on the metal-non metal transition and conductivity of Fe_3O_4 *Phys.Rev.Lett.* **21** 795–7
[18].Prakash R, Choudhary R J, Sharath Chandra L S, Lakshmi N and Phase D M 2007 Electrical and magnetic transport properties of Fe_3O_4 thin films on a GaAs(100) substrate *J. Phys. Condens. Matter***19** 0953-8985
[19]. Khan M A M, Khan W, Ahamed M and Alhazaa A N 2019 Investigation on the structure and physical properties of Fe_3O_4/RGO nanocomposites and their photocatalytic application *Mater. Sci. Semicond. Process.* 99 44–53

Acknowledgments

RSJ acknowledges Vision Group of Science and Technology, Government of Karnataka for Young Scientist award, UGC BSR start-up grant and Central University of Karnataka for the financial support to carry out this work.

IOP Conf. Series: Materials Science and Engineering 1124 (2021) 012003 doi:10.1088/1757-899X/1124/1/012003

High speed laser induced periodic structure generation using femtosecond laser on surface of titanium

A Mundamajhi[1], S K S Parashar[2], S K Das[2*]
1 Dept. of Physics, School of Applied Sciences, Kalinga Institute Of Industrial Technology, Deemed to be University, Bhubaneswar,Odisha, 751024, India
2 School of Physics, Gangadhar Meher University, Amruta Vihar, Sambalpur, Odisha, 768004, India

Abstract. In this work femtosecond laser wavelength of 800nm,1KHZ repetition rate and pulse length 100fs irradiation surface texturing has been used.The laser surface texturing is distinguished itself than other conventional processing due to its simplicity controllability and reproducible in creating various structures and wide range of applications .A labview compact zaber xy stage is used for this purpose and investgated the high scanning speed (20mm/s,10mm/s) surface structures on Titanium plate.The velocity of sample position used as function to generate modified surface structures. The surface textured studied by effect of variation of scanning velocity and this texturing process has been done in air medium by above selected parameters. FESEM sample characterization study of surface morphology has been done to details study of textured sample surface..

1. Introduction

In the current era Titanium is sought after material being widely implements in biomedical applications. This material shows excellent chemical and physical properties unlike high erosion resistance, high melting point [1] .The main application of Titanium implants in medicine is based on fact that it posses high level of bio integration and bio compatibility with human body [2-9]. The young's modulus of Titanium is nearer to human bone. So it is mostly used in joint replacement part for knee, wrist, hip, spine, shoulder etc. In other field this implant can be successfully used for replacement of lost teeth[3, 5, 6, 9]. To increase the reliable growth of cell over Ti implant surface texturing is generally done. The different methods of surface texturing are ion implantation sandblasting, hydrothermal modification, surface coating, alkaliheat treatment, plasma-spraying and acid etching [10].Laser processing is another method to texture the surface any solid-state material. This method is generally very reproducible and controlable. Among various types of lasers, the fs laser generally can give best result because of its non-thermal nature of material processing. Various works have been done on texturing of Ti using fs laser.HUANG et.al demonstrated 2d periodic structures using femtosecond laser[11-12].Cunha et.al performed texturing Ti surface by using Yb:KYW chirped pulse fs laser for for reduction adhesion of staphylococcus aureus and biofilm formation[13]. Kudryasove et.al demonstrated Nanostructuring of solid surfaces by femtosecond (fs) laserpulses [14]. Ionin et.al demonstrated Sub-100 nanometer transverse gratings generation by femtosecond (fs) pulses on a surface of titanium [15]. Bonse et.al demonstrated sub-100nm periods of laser induced periodic structures [16-17]. Wang et.al studied the pulse energy threshold for surface modification with femtosecond laser pulse irradiations [18].Nayak et.al demonstrated Formation of

[1]rajeevsj@cuk.ac.in

nano-textured conical microstructures in titanium metal surface by femtosecond laser irradiation[19]. Golosove et.al Femtosecond Laser induced Subwave One-Dimensional Quasiperiodic Nanostructures [20] .Vorobyev et.al has done texturing of Ti implant using fs laser [21]. Tsukamoto et.al demonstrated the Microstructures formation on titanium plate by femtosecond laser ablation[22].Zhidkov et.al demonstrated the generation of the oxide phases on the surface of titanium using multipulse femtosecond laser irradiation[23].Shinonago et.al demonstrated the growth of periodic nanostructures on titanium surface by femtosecond laser and control of cell spreading on it[24].Li et.al demonstrated the femtosecond laser induced 100-nm periodic structures by exploiting the oxidation and third harmonic generation on a titanium surface [25]. For large area texturing generation, roster scanning of the sample is done across the focus of the laser. In order to reduce the processing time, in the fs laser texturing of Ti surface in the work is done at high speed. In this present work the effect of variation of sample velocity on morphology of the generated structures on Ti surface is also investigated.

2. Experimental

A femtosecond laser wavelength of 800nm and pulse length 100fs irradiation was used for surface texturing Ti. The laser beam was focused onto the sample by using a lens of focal length of 5 cm. The sample was placed on a computer controlled stage to move it at the desired speed across the focus. The experiments were carried out at three different velocity 20 mm/s and 10mm/s . For all the cases the fluence was kept at 0.5J/cm^2. All the texturing process was done in air medium. Field Emission Scanning Electron Microscopy (FESEM) was done to find the morphology of the generated surface structures.

3. Results and Discussion

The FESEM image of the generated structures for 20 mm/s is shown in Figure. 1. From this just appearance of the laser induced periodic structures (LIPSS) was found. The aforementioned fluence was just equal to the laser damage threshold and hence the LIPSS were not found to be so distinct.
On the other hand for the velocity of 10 mm/s, quite distinct and uniform LIPS were found (Figure 2a). With the decrease of the scanning speed generally the LIPSS laser damage threshold decreases. So the aforementioned is little more than the laser damage threshold. It has been already reported that the distinct and uniform LIPS can be generated if the fluence is little more than the damage threshold. In our scanning case of 10 mm/s this condition has been matched and hence we get the distinct and uniform LIPS on Ti surface. To know the further details the two dimensional Fast Fourier transformation (2DFFT) analysis was done. From this 2DFFT analysis (Figure. 2b) the periods of these structures were 600 nm.

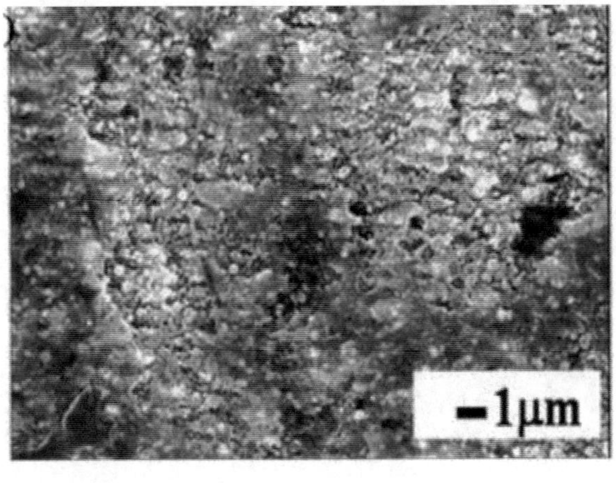

Figure1.FESEM images of textured Ti surface for scanning speed 20mm/s.

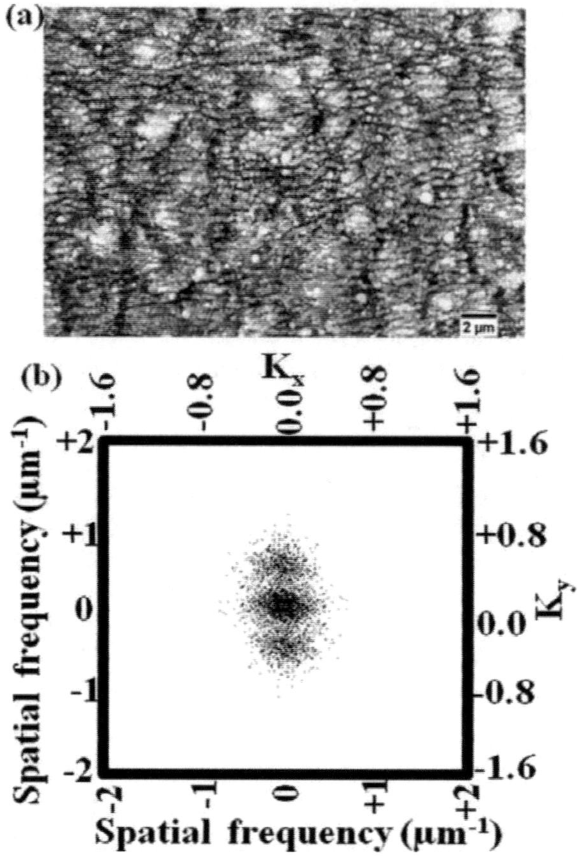

Figure.2. (a) FESEM images of LIPS generated on Ti surface with the sample scanning speed of 10mm/s (b) its 2DFFT diagram

Table I: Scanning speed reported so far for generation of LIPSS in large surface area of Ti using fs laser are given in Table I.

Detail of fs Laser used (Pulses duration τ, central wavelength λ)	Scanning speed	Reference
80 fs, 744 nm	20μm/s	[14]
80 fs, 744 nm	20μm/s	[20]
150 fs, 775 nm	2mm/s	[24]
100 fs, 800 nm	1mm/s	[25]
30 fs, 790 nm	1mm/s	[26]
550 fs, 800 nm	0.2mm/s - 2.5mm/s	[27]
100 fs, 800 nm	20 mm/s	Current work

From this it clear that scanning speed reported by us is the highest so far. This makes the LIPS generation work on Ti surface very less consuming and hence more suitable for the aforementioned applications.

4. Conclusion

The laser surface texturing is distinguished itself than other conventional processing due to its simplicity controllability and reproducible in creating various structures and wide range of applications .A labview compact zaber xy stage is used for this purpose and investgated the high scanning speed (20mm/s,10mm/s) surface structures on Titanium plate. The velocity of sample position used as function to generate modified surface structures. The surface textured studied by effect of variation of scanning velocity and this texturing process has been done in air medium by above selected parameters.

5. References

[1] Goodfellow Catalogue, Goodfellow Cambridge Ltd. England, (2000).

[2] Simpson J.p, Geret V, Brown S.A,Merritt K, "Retrieved fracture plates-implant and issue analysis," in Implant retrieval-material and biological analysis, A. Weinstein, D.Gibbons, S. Brown and S.Ruoff, eds. *NBS Spec.Publ.* **601** (1981), 395

[3] Steinemann S.G, Metal implants and surface reactions, *Injury*, **27**(1996) (Supl. 3) SC16.

[4] Long M, Rack H.J, Review Titanium alloys in total joint replacement—a materials science perspective, *Biomaterials*, **19** (1998) 1621.

[5] Peto G, Karacs A, Paszti Z, Guczi L, Divinyi T, Joob A, Surface treatment of screw shaped titanium dental implants by high intensity laser pulses,*Appl. Surface Sci.* **186** (2002)7.

[6] Bereznai M, Pelsoczi I,Toth Z, Turzo K,Radnai M, Bor Z, Fazekas A, Surface modifications induced by ns and sub-ps excimer laser pulses on titanium implant material, *Biomaterials*, **24** (2003) 4197.

[7] Guillemot F, Prima F, Tokarev V.N, Belin C,Porte-Durrieu M.C, Gloriant T, Baquey C, Lazare S, Single-pulse KrF laser ablation and nanopatterning in vacuum of β-titanium alloys used in biomedical applications,*Appl.Phys. A*, **79** (2004) 811.

[8] Lima M.S.F, Folio F, Mischler S, Microstructure and surface properties of laser-remelted titanium nitride coatings on titanium Surface & Coat. *Technology*, **199** (2005) 83.

[9] Kim S.G,Kim M.D, Eom T.G, Yoon J.H, Sang-Gun A. Surface Properties of Endosseous Dental Implants After NdYAG and CO2 Laser Treatment at Various Energies, *J. Oral Maxillfac. Surg.* **63** (2005) 1522. 10

[10] John A.A, Jaganathan S.K, Supriyanto E and Manikandan A, Surface modification of titanium and its alloys for the enhancement of osseointegration in orthopaedics, *CURRENT SCIENCE*, **111** (2016) 6,

[11] Huang Y, Liu S, Li W, Liu Y, and Yang W,Two-dimensional periodic structure induced by single-beam femtosecond laser pulses irradiating titanium, *Optics Express*, **17**, (2009)23pp. 20756-20761

[12] Perez del Pino A, Serra P, Morenza J.L, Coloring of titanium by pulsed laser processing in air, *Thin Solid Films,* **415** (2002) 201–205

[13] Elieb A.M,Plawinski L,Serrod A.P,Botelho do Regoe A.M, Almeida A,Urdacib M.C,Durrieu M.C,Vilara R,Femtosecond laser surface texturing of titanium as a method toreduce the adhesion of Staphylococcus aureus and biofilm formation , *Applied Surface Science*, **360** (2016) 485–493

[14] Kudryashov S.I, Golosov E.V, Ionin A. A, Kolobov Y.R, Ligachev A.E,Seleznev L.V,Sinitsyn D.V, Sharipov A.R, Nanostructuring of solid surfaces by femtosecond laser pulses , *AIP Conf. Proc.* **1278** (2010) 156

[15] Ionin A.A,Kudryashov S.I,Makarov S.V,Seleznev L.V, Sinitsyn D.V,Ligachev A.E,Golosov E.V and Kolobov Y.R,Sub-100 nanometer transverse gratings written by femtosecond laser pulses on a titanium surface, *Laser Phys. Lett.* **10** (2013) 056004 (5pp)

[16] Poua P, Riveiro A, del Vala J, Comesaña R, Penidea J,Arias-Gonzáleza F, Sotoa R , Flusquinos pou J, Laser surface texturing of Titanium for bioengineering application,*Procedia Manufacturing*, **13** (2017) 694–701

[17] Bonse J, Höhm S,Rosenfeld A,Krüger J, Sub-100-nm laser-induced periodic surface structures upon irradiation of titanium by Ti:sapphire femtosecond laser pulses in air, *Appl Phys: A,* **110** (2013) 547–551
[18] Wang C, Huo H,Johnson M,Shen M and Mazur E,Thresholds of surface nano-/micro-morphology modifications with femtosecond laser pulse irradiations, *Nanotechnology*, **21** (2010) 075304 (5pp)
[19] Nayak B.K,gupta M.C, kolasinski K.W, Formation of nano-textured conical microstructures in titanium metal surface by femtosecond laser irradiation, *Appl.Phys.A,* **90** (2008)399–402
[20] Golosov E.V, Emelyanov V.I, Ionin A.A, Kolobov Y.R", Kudryashov S.I,Ligachev A.E, Novoselov Y.N, Seleznev L.V, and Sinitsync D.V ,Femtosecond Laser Writing of Subwave One-Dimensional Quasiperiodic Nanostructures on a Titanium Surface, *JETP LETTERS,* **90** (2009) 2
[21] Vorobyev A.Y,Guo C, Femtosecond laser structuring of titanium implants, *Appl. Surf.Sci.*, **253** (2007) 7272–7280
[22] Tsukamoto M,Kayahara T, Nakano H, Hashida M, Katto M, Fujita M, Tanaka M and Abe N, Microstructures formation on titanium plate by femtosecond laser ablation, *Journal of Physics: Conference Series*, **59** (2007) 666–669
[23] Zhidkov M.V, Golosov E.V,Vershinina T.N, Kudryashov S.I, Kolobov Y.R and Ligachev A.E,Formation of the oxide coating on the titanium surface by multipulse femtosecond laser irradiation, *Journal of Physics: Conf. Series* **1115** (2018) 042066
[24] Shinonaga T, Tsukamoto M,Kawa T, Chen P, Nagai A, Hanawa T,Formation of periodic nanostructures using a femtosecond laser to control cell spreading on titanium, *Appl. Phys. B*,**119** (2015)493–496
[25] Li X.F, Zhang C.Y,Li H, Feng Dai Q, Lan S, and Tie S,Formation of 100-nm periodic structures on a titanium surface by exploiting the oxidation and third harmonic generation induced by femtosecond laser pulses, *OPTICS EXPRESS* , **22** (2014) No. 23
[26] Bonse J, Höhm S, Koter R, Hartelt M, Spaltmann D, Pentzien S, Rosenfeld.A, Krüger J, Tribological performance of sub-100-nm femtosecond laser-induced periodic surface structures on titanium, *Appl. Surf.Sci.* **374** (2016) 190–196
[27] Nathala C.S.R, Ajami A, Ionin A.A,Kudryashov S.I, Makarov S.V, Ganz T, Assion A, Husinsky W, Experimental study of fs-laser induced sub-100- nm periodic surface structures on titanium, *OPTICS EXPRESS* **5916**(2015) Vol. 23, No. 5
[28] Nowak R, Metev S, Thermochemical laser etching of stainless steel and titanium in liquids, *A. Phys. A* ,**63**, 133-138 (1996)

IOP Conf. Series: Materials Science and Engineering 1124 (2021) 012004 doi:10.1088/1757-899X/1124/1/012004

Influence of Thiourea Metal Complex on Potassium Dihydrogen Phosphate for Optoelectronic Applications

Y B Rasal[1]

[1]Smt. S. K. Gandhi Arts, Amolak Science and P. H. Gandhi Commerce College Kada, Dist. Beed, Maharashtra, India

Abstract. Influence of 0.1 mole % Bis Thiourea Zinc Sulphate (BTZS) exhibits concrete reinforcing in optical and dielectric properties of Potassium Dihydrogen Phosphate (KDP). The pure and BTZS influenced KDP crystals have been grown by slow evaporation solution technique at room temperature. The structural and vibrational studies of the grown crystals have been studied by single crystal XRD and FT-IR technique respectively. The optical traits of the doped crystal have been analyzed by UV-vis spectral analysis within the wavelength range of 200-800 nm. The Kurtz-Perry powder test has been employed to determine the bisthiourea zinc sulphate on the second harmonic generation (SHG) efficiency of KDP crystal and it found to be 1.1 times higher than KDP. The dielectric and thermal behaviour has been analyzed for doped KDP crystal.

1. Introduction

The researcher and academicians were showing interest in the non linear optical (NLO) material from last few decades due to its fascinating materialistic properties. The NLO crystals study involves the interaction of light with the crystal resulted in the variation in the electromagnetic mechanism of the incident light. The most fascinating NLO applications of organic, inorganic and semi-organic crystals have a usage in the optical traits. These traits are optical storage, optical computing, optical information processing, optical power limiting, optical switching, antireflection coating, image manipulation and processes. The rapidly growing field in laser technologies, frequency conversion devices, holographic memory, electro-optics modulation and photonics by second order effects in crystals has fulfilled many more needs of today's industry and society [1-4]. The organometallic crystals have superior optical, dielectric and mechanical properties. The large dipole moments, the ability to form metal ligands through hydrogen bonding effectively works to improve optical properties [5]. The optical properties of the subjected crystal enhanced during chemical reaction and due to the bonding of the thiourea and zinc sulphate. The donor and acceptor electron system and formation of the hydrogen bonding throughout the crystal enhances second harmonic efficiency due to engagement of centrosymmetric thiourea compound and metal component [6]. The thiourea metal complexes have applications in optical phenomenon due to its low UV cut off wavelength and high NLO properties. In Literature, the NLO zinc thiourea sulphate (ZTS) is found to be a versatile organometallic crystal [7]. The potassium thiourea bromide (PTB), bis-thiourea cadmium acetate (BTCA), copper thiourea chloride (CTC), bis-thiourea zinc acetate (BTZA), zinc thiourea chloride (ZTC), bis-thiourea cadmium chloride (BTCC) are some thiourea metal complexes found in the

[1]rasyog1975@gmail.com

literature [8].The fundamental material in the NLO crystal category is potassium dihydrogen phosphate (KDP) due to its unique nature and properties. The different doping attempts were made into KDP crystal to improve various properties [9-10].
The enhanced SHG efficiency reported doped thiourea complexes in KDP are bis thiourea Copper complex [11], tetra thiourea Potassium chloride [12], thiourea Ammonium chloride [13] and bis thiourea zinc chloride [14]. In current research an investigation of BTZS in KDP is put forward, by imposing Single crystal XRD, FT-IR, UV-visible, Kurtz-Perry test, thermal and dielectric analysis to bring out the potential credibility in NLO device applications.

2. Experimental procedure

To synthesize the thiourea (Bis) Zinc Sulpahte doped KDP (BTZSKDP) complex starting material thiourea and zinc sulphate were dissolved in double distilled water whose conductivity is less than 1.0 μmhos in 2:1 M ratio for to produce thiourea (Bis) Zinc Sulpahte. After stirring the mixture for four hours homogeneous solution was filtered and kept for slow evaporation. The BTZS crystal was found according to reaction

$$2CS(NH_2)_2 + ZnSO_4 \longrightarrow Zn[CS(NH_2)_2]_2SO_4$$

The good yield of thiourea (Bis) Zinc Sulpahte (BTZS) was collected and made fine microelements for further crystal growth. The precisely measured 0.1 M% and 0.2 M% BTZS was added gradually in the distinct beakers of supersaturated solution of the Potassium dihydrogen phosphate (KDP) and allowed to stir at a constant speed. The solution was constantly stirred for five hours to prepare the homogenous mixture of added reactants. The solution was filtered through 4μm membrane filter paper and allowed to place in the clean rinsed beaker in a vibration-free atmosphere at room temperature. During the recrystallization process purity enhanced resulting in good quality crystals within a period of two weeks. The 0.1 M % BTZSKDP crystal is as shown in figure 1; also Table 1 shows the optimized growth mechanism 0.1 of BTZSKDP single crystal.

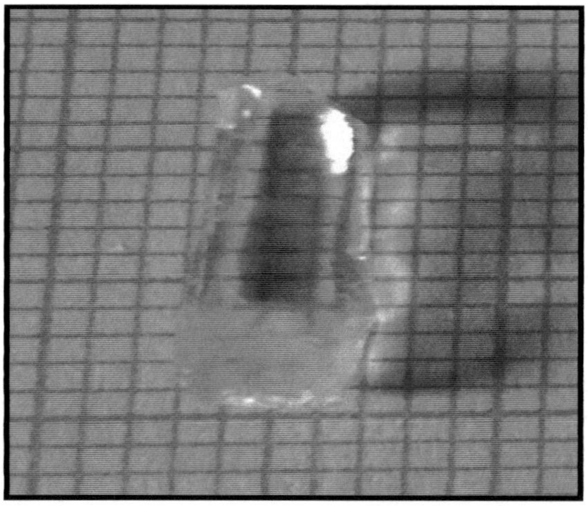

Figure 1. Grown crystal of BTZSKDP crystal

3. Results and discussion

3.1. Single crystal x-ray diffraction studies

The well-phased crystal of 0.1 BTZSKDP was subjected to single crystal x-ray diffraction studies. The study was carried out using Enraf Nonius CAD4-MV31 crystal X-ray diffractometer. The analysis of single crystal XRD of BTZSKDP confirms the tetragonal I crystal system. The determined lattice parameter values are a = 7.44 Å, b = 7.44 Å, c = 6.96 Å and $\alpha = 90^0$, $\beta = 90^0$, $\gamma = 90^0$ with volume 386 $(Å)^3$.

3.2. Fourier transform infrared (FT-IR) spectral analysis

The incorporation of bis-thiourea zinc sulphate in KDP was qualitatively analyzed by recording the FT-IR spectrum for doped crystal by using Bruker α-ATR spectrophotometer. Figure 2 shows the plot of FT-IR spectrum; was recorded in the wave number range between 500 cm^{-1} to 4000 cm^{-1}. The N-C-N bond stretching vibration associated with ZTS is observed at 570 cm^{-1}.The absorption peak observed at 624 cm^{-1} corresponds to the C=S stretching vibration. The P=O and SO_3 - stretching is evidenced at wave number 1090 cm^{-1}.The O=P–OH stretching absorption peak are observed at 1518 cm^{-1} in doped The C=N-OH bond stretching vibration is observed at 1717 cm^{-1}. The strong P-H stretching vibration is attributed to 2349 cm^{-1}. The absorption peak at 3070 confirms the N-H stretching vibration implies that the presence of thiourea molecule in the doped crystal. The peaks between 3732 to 4000 cm^{-1} appears due to O-H and N-H bond stretching vibrations [15].

Table 1. Optimized growth mechanism of BTZSKDP.

BTZSKDP crystal	Parameter
Solvent used	Double distilled water
Molar ratio of thiourea and Zinc Sulphate	2:1
Molar ratio of BTZS in KDP	0.1 and 0.2
Size of grown crystal	1.2 × 0.5 × 0.4 cm^3
Time of growth	2 weeks
Volume of crystal	386 $(Å)^3$

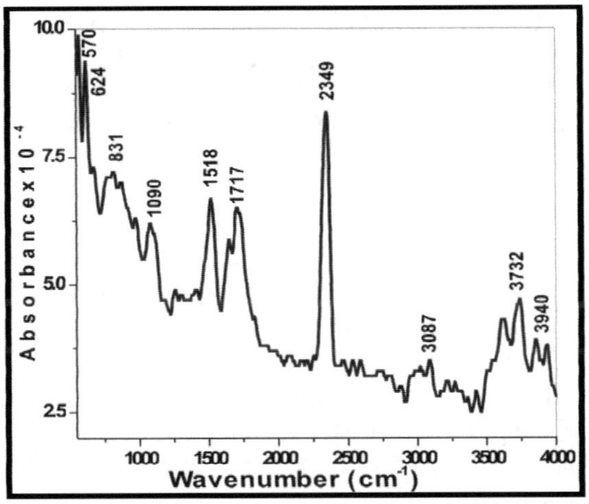

Figure 2. FT-IR spectrum of doped KDP crystal

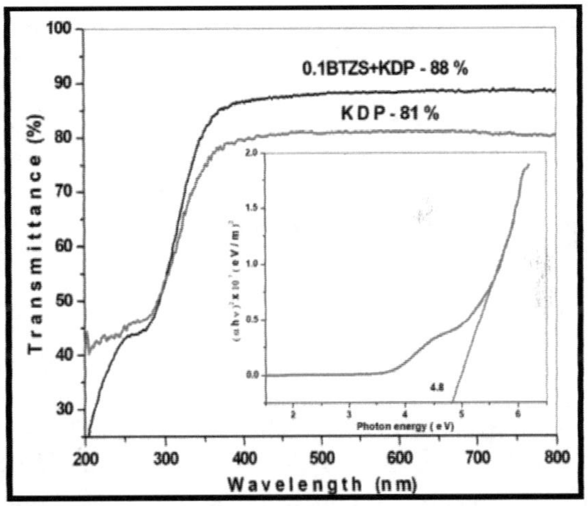

Figure 3. Plot of a) Wavelength vs. transmittance; b) Tauc's plot

3.3 SHG Studies

The Kurtz-Perry powder technique is used to study the frequency doubling of the 0.1 and 0.2 M % BTZSKDP and KDP crystals owing to determine frequency conversion efficiency [16]. The fundamental Q-switched Nd:YAG laser beam of wavelength 1064 nm producing pulse width 6 ns with a repetition rate of 10 Hz interact with the subjected crystal. The good quality crystals of 0.1, 0.2 M % BTZSKDP and KDP converts into uniform microgrannuals and sieved in micro-capillary and was exposed by a beam of Pulse energy 4.7μJ/pulse. The output signals emitted from the sample was focused on monochromator using a pair of lenses and collected by photomultiplier tube were displayed on CRO. The emission of the sharp green radiation of wavelength 532 nm confirms the second order nonlinearity of the subjected crystals. The output voltages of the second order diffracted radiation beams were recorded and found the values 124 mV, 84 mV and 114 mV for 0.1, 0.2 M % BTZSKDP and KDP respectively. This shows that the 0.1 M % BTZSKDP have remarkable enhancement in the SHG efficiency than parent material is due to higher polarizing ability, more electrons-phonon interaction, noncentrosymmetric nature of crystal and enhanced charge transfer through organometallic BTZS complex [17-18]. Table 2 shows a comparative chart of the SHG efficiency of doped KDP crystals with KDP [7]. The 0.1 M % ZTSKDP crystals may show better alternative results in the frequency conversion and optoelectronic applications [17-19].

Table 2. Comparative chart of SHG efficiency.

Doped KDP Crystal	SHG Efficiency	Ref.
CuT	1.2	[11]
PTTC	1.24	[12]
TAA	1.27	[13]
BTZC	1.65	[14]
0.1 M % BTZS	1.1	[Present]
0.2 M % BTZS	0.7	[Present]

3.4 UV-visible transmission studies

The well-polished 2 mm thick crystal of 0.1 M % BTZSKDP crystal was subjected for UV-visible studies by using Shimadzu UV-2450 spectrophotometer. The crystal was placed in the holder and allowed to pass UV light of the wavelength 200 nm to 800 nm range. The recorded spectrum of transmittance as a function of wavelength was depicted in figure 3 (a) and subjected crystal possesses optically transparency up to 88 % in the entire visible region and lower cut off wavelength at 205 nm. And figure 3(b) shows Tauc's plot and confirms the band gap energy for the exposed crystal as 4.8 eV. Table 3 shows a comparative chart of cut off wavelength of the doped and undoped KDP crystals. The higher value of transmittance and cut off wavelength strongly agree for the suitability of the crystal in non linear optics for frequency conversion device applications.

Determination of optical parameters:

The optical properties of the crystal relate to the atomic structure, electronic band structure and electrical behaviour. The various optical parameter values are prerequisites for the selection of the crystal in non linear, optoelectronic applications and in device fabrication [20]. In the present study different optical parameters such as optical transmittance (T in %), band gap energy (E_g in eV), cut off wavelength (λ_{co}in nm), absorption coefficient (α), reflectance (R), refractive index (n), optical conductivity (σ in S^{-1}) and extinction coefficient (K) were calculated by using recorded optical transmission spectrum. The magnitude of transmittance values and cut off wavelength has already discussed and other optical parameters were determined as below.

Absorption coefficient and band gap energy:

To determine the optical band gap initially the optical absorption coefficient was calculated from the transmittance data by using relation

$$\alpha = \frac{1}{d}\text{Ln}\frac{1}{T}, \quad (1)$$

Where, T is the transmittance and d is thickness of the crystal. The Tauc's plot shows the dependence of absorption coefficient (α) on the incident energy of photons (E_g) by the relation

$$\alpha h\nu = A(h\nu - Eg) \quad (2)$$

Where, E_g is the optical band gap energy and A is constant. The photon energy E_g value is determined at an extrapolating onto the photon energy axis in the linear portion near the onset of the absorption edge as plotted in figure 3 (b) Tauc's plot and found to be 4.8 eV. This reflects the suitability in of the subjected crystal optoelectronic applications [21].

Refractive Index and Reflectance:

The propagation of light through the medium of the material is a refractive index as shown in figure 4(a) and was determined by using the above formula and found to be 1.5 in the entire visible region.

$$n = [\frac{1}{T} + (\frac{1}{T} - 1)] \tag{3}$$

Similarly, reflectance in terms of refractive index was evaluated by the relation

$$R = (n-1)^2/(n+1)^2 \tag{4}$$

And, which is observed from the figure 4(b) has value 0.05 % in the entire visible region. The high value of transmittance and band gap energy and lower values of reflectance of BTZSKDP in the entire UV-vis region plays a crucible role in the antireflection coating in solar thermal devices [22].

Table 3. Comparative chart of cut off wavelength.

Doped KDP Crystal	Cut off Wavelength (λ_{CO})	Reference
PTTC	299	[12]
BTZC	300	[14]
BTNN	299	[20]
BTZS	205	[present]

Optical conductivity and Extinction coefficient:

The optical conductivity was calculated by using formula $\sigma = \frac{\propto nC}{4\pi}$, where, α is the absorption coefficient, n is refractive index and C is velocity of light and response to the photon energy is plotted in figure 5(a). The optical conductivity confirms the presence of high photo response of the crystal [23]. This shows doped crystal has good matching properties for the use in optical information and processing purpose. The formula $k = \frac{\propto\lambda}{4\pi}$ is used to determine the extinction coefficient and its response to wavelength is depicted in figure 5(b) and shows that it depends on wavelength. The lower extinction coefficient facilitates less absorption of photons; vital for UV tunable laser, telecommunication applications [23-24]

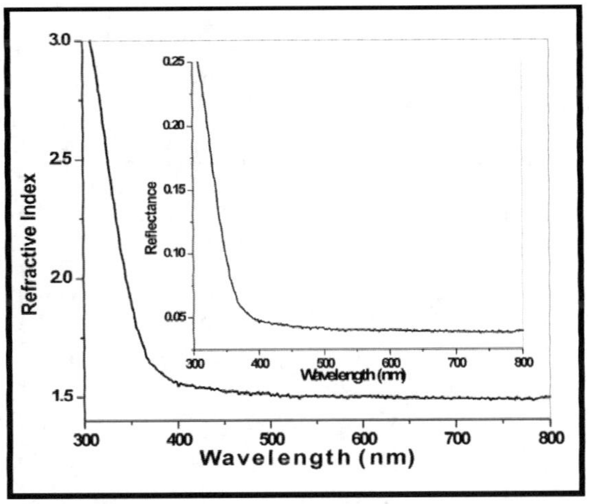

Figure 4. Plot of wavelength vs.
a) refractive index; b) reflectance

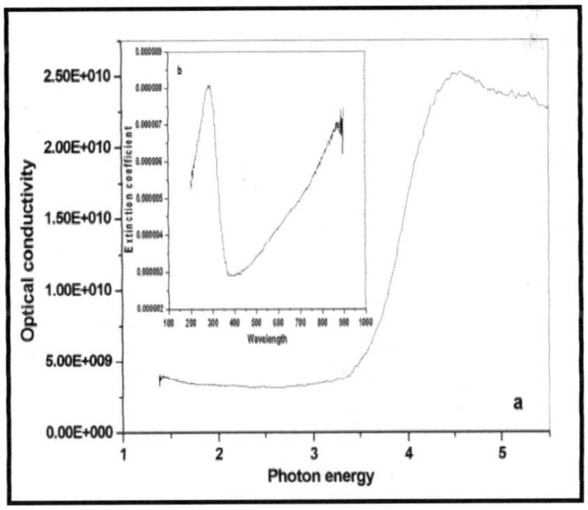

Figure 5. Plot of photon energy vs.
a) opt. conductivity; b) ext. coefficient

3.5. Thermal Studies

The thermogram of the 0.1 BTZSKDP crystal was used to study response to the temperature variations using Thermogravimetric (TG) and Differential Thermal (DTA) using detector DTG-60H in the temperature range of 25-500 ^{0}C at a heating rate of 10 °C/min in a nitrogen atmosphere. The well phased good quality crystal of doped crystal was crushed into powder form subjected for thermal study. The sample weight used; 5.250 mg during the TGA process. The thermogram was illustrated in figure 6 and it is observed that decomposition of the sample starts at 65 ^{0}C and ends at 450 ^{0}C. During this process, 0.836 mg weight loss of the sample takes place which is around 15.9 % determined from the TGA curve. The major weight loss is produced from 208 ^{0}C to 352 ^{0}C. Also, it is confirmed from DTA curve initial loss at 229 ^{0}C, the sharp onset of the first endothermic fall occurs due to corresponding weak forces in the molecules and which is melting point of the doped material. The total decomposition of the material produced into another two stages at 266 ^{0}C and 305 ^{0}C. It is concluded from the thermal study that the ZTSKDP crystal has optimum thermal stability and may have applications in NLO and optoelectronic devices up to 229 ^{0}C temperature [13, 25].

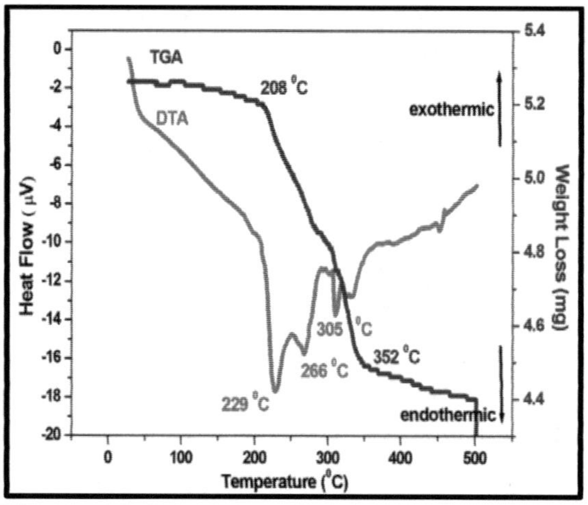

Figure 6. TGA-DTA analysis of 0.1 BTZSKDP

3.6. Dielectric Studies

The good quality crystal of 0.1 M % BTZSKDP was subjected to study dielectric behaviour using Gwinstek LCR-819 meter. The measurements of capacitance were taken for the frequency range of 12 Hz to100 kHz for the subjected crystal at room temperature.

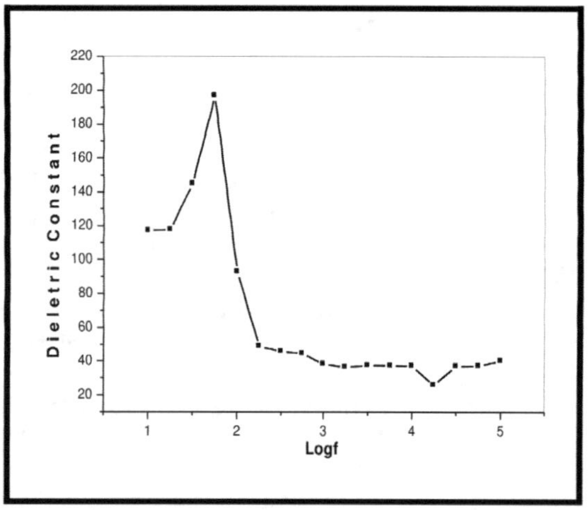

Figure 7. Plot of Log f vs. dielectric constant

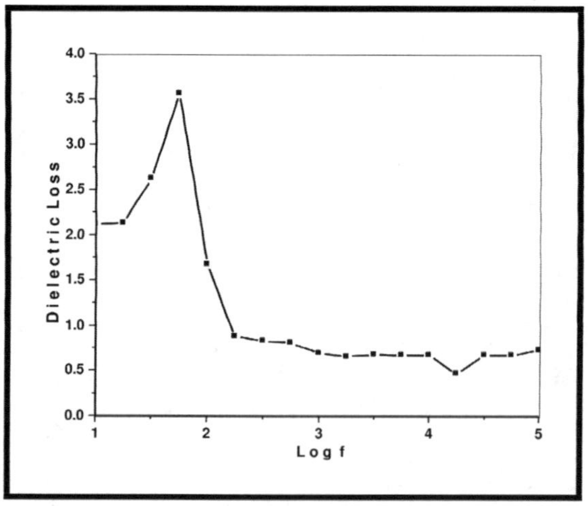

Figure 8. Plot of Log f vs. dielectric Loss

Dielectric constant

The recorded data was used to calculate the dielectric constant (DC) of the subjected material by using formula

$$\varepsilon = \mathrm{Cd} / \varepsilon_o \mathrm{A} \quad (5)$$

Where C is the capacitance, d is the thickness and A is the area of crystal sample.

The response of the dielectric constant to the frequency was depicted in the figure 7 and decay observed in a lower frequency region above 100 Hz. The doping of thiourea metal complex makes superior quality than pure KDP. The DC value is mainly attributed to the charge delocalization among the crystal contributes to the frequency dependent electronic, ionic, dipolar space-charge polarizations. And the purity and lattice perfection of the crystal affects space charge polarization which is generally present in the low frequency domain [26].

Dielectric loss

The dielectric loss (DL) response to the frequency of the subjected crystal depicts in figure 8 and determined by using formula

$$\text{Dielectric loss } (\mathrm{Tan}\ \delta) = \text{Dissipation factor} \quad (6)$$

The dissipation factor is measured using LCR meter along with the capacitance varied with frequency at room temperature. The DL is the dissipation of electromagnetic energy due to the heating of dielectric material when applied to the frequency dependent field. It can be parameterized in terms of either the loss angle δ or the corresponding loss tangent (Tan δ). The behaviour of the DL at lower frequencies have decaying response to the applied field cognates that there are less defects produced in the KDP crystal caused by doping [27]. The enhanced second order nonlinearity and optical parameters were assigned to lower magnitude of the dielectrics. This Miller theory holds well in the BTZSKDP crystal and founds its suitability in the frequency conversion and photonic devices [14, 28-30].

4. Conclusions

The optical quality 0.1 M% BTZSKDP crystal was successfully grown by slow evaporation solution technique. The single crystal XRD analysis confirmed the tetragonal structure. The presence of functional groups was confirmed by FTIR analysis. The enhancement in optical transmittance of KDP found due to the doping of ZTS. The high value of band gap energy and lower values of reflectance and refractive index of grown crystal plays crucible role in the antireflection coating in solar thermal devices. The SHG efficiency of 0.1 M % BTZSKDP crystal is enhanced than KDP crystal and found to be 1.1 times that of KDP. The thermal study reveals that the material is suited up to 229 ^{0}C for applications. The dielectric constant and dielectric loss was dropped to optimum value due to the presence of BTZS. The 0.1 M % BTZSKDP crystal with impressive optical and dielectric properties might find suitable applications in designing distinct NLO and photonic devices.

References

[1] Ambrose R M, Stanly J X S, Anbarasu S and Prem A D 2017 *Mater. Rese. Innova.* **23,** 73
[2] Mohanraj V, Sakthivel P, Ponnuswamy S, Muthuraja P and Dhandapani M, 2019 *Mater. Today* **8** 1
[3] Mahadevan M, Magesh M, Ramachandran K, Anandan P, Arivanandhan M and Hayakawa Y 2014 *Spectroc. Acta,* **130,** 416
[4] Kulkarni R B, Anis M, Hussaini S S and Shirsat M D 2019 *Modern Phys. Lett. B,* **1850424** - 1
[5] EzhilVizhi R, Kalainathan S, and BaghavanNarayana G 2007 *Cryst. Res. Technol.* **42,** 1104
[6] Jagadeesh M R, Suresh Kumar H M and Ananda K. R. 2015 *Optik* **15,** 00743
[7] Marcy H 0, Warren L F, Webb M S, Ebbers C A, Velsko S P, Kennedy G C and Catella G C 1992 *Applied Optics* **31,** 5051
[8] Mohd A., Hussaini S S, Hakeem A, Shirsat M D and Muley G G 2015 *Optik* **127,** 2137
[9] Dolzhenkova E F, Kostenyukova E I, Bezkrovnaya O N and Pritula I M 2017 *J. Crystal Growth* **17,** 30493

[10] Kubendiran T, Ravi K S M, Allen Moses S E, Nasareena B. A, Shanthi C and Sivaraj S 2019 *J. Mater. Sci. Mater. Electr.* **30,** 17559
[11] Kumaresan P, MoorthyBabu S and Anbarasan P M 2007 J. *Optoelect. Adv. Mater.* **9,** 2787
[12] Rasal Y B, Shaikh R N, Shirsat M D, Kalainathan S, and Hussaini S S 2017 *Ferroelectrics* **520,** 59
[13] Rasal Y B, Kulkarni R B, Shirsat M D and Hussaini S S 2019 *Ferroelectrics* **537,** 37
[14] Rasal Y B, Anis M, Shirsat M D and Hussaini S S 2017 *Mater. Rese. Innov.* **21,** 45
[15] Colthup N B, Daly L H and Wiberley S E 1990 *Infrared Spectroscopy, American Cyanamid Company San Diego, California* Toronto: Academic Press, Print Book ISBN: 9780121825546.
[16] Kurtz S K and Perry T T 1968 *J. Appl. Phy.* **39,** 3798
[17] Sathiskumar S 2019 *J. Crystal Growth* **526,** 125234
[18] Ramteke S P, Baig M I, Mohd S., Kalainathan S, Shirsat M D, Muley G G and Mohd A. 2018 *Optics Laser Technol.* **104,** 83
[19] Rasal Y B, Shaikh R N, Shirsat M D, Kalainathan S and Hussaini S S 2017 *Mater. Res. Express* **4,** 036202
[20] Balaji J, Srinivasan P, Prabu S, Merin G and Sajan D 2020 *J. Molecul. Structure* **1207,** 1
[21] Senthil K, Kalainathan S, Hamada F, Yamada M and Aravindan P G 2015 *Optic. Mater.* **46,** 565
[22] Ramteke S P, Mohd A., Baig M I, Pahurkar V G and Muley G G 2017 *optik* **137,** 31
[23] Sabari G T C and Dhanuskodi 2009 S *Cryst. Res. Technol.* **44,** 1297
[24] Subashini A, Rajarajan K, Sagadevan S, Singh P and Podder J 2017 *Therm J. Anal Calorim* doi 10.1007/s10973-017-6829-8.
[25] Thangaraj M, Ravi G, Sabari G.T.C., Vinitha G and Loganathan A, 2014, **138,** 158
[26] Senthil P M and Ramasamy P 2010 *J. Crystal Growth* **312,** 413
[27] Lunkenheimer P, Krohns S, Gemander F, Schmah W W and Loid A 2020 *Scient. Repor.* **4** 6020
[28] Kalaiselvi D and Jayavel R 2012 *Appl Phys A* **107,** 93
[29] Mohd A, Shirsat M D, Muley G and Hussaini S S 2014 *Physica B* **449,** 61
[30] Shaikh R N, Shirsat M D, Koinkar P M and Hussaini S S 2015 *Optics Laser Technol.* **69,** 8

Dielectric, Microhardness and Thermal Properties of Swift Ion (Au^{3+}) Irradiated NLO Single Crystal: 2-Amino-5-Nitropyridinium Sulfamate (2A5NPS)

M. Ambrose Rajkumar[1*], P. Ajith[2], S. Anbarasu[3], Prem Anand Devarajan[2]

1. Department of Physics, Kristu Jayanti College (Autonomous), Bengaluru-560077, India
2. Materials Research Centre, Department of Physics, St. Xavier's College (Autonomous), Palayamkottai, Tamilnadu-627002
3. Department of Physics, St.Joseph's College (Autonomous), Tiruchirappalli-620 002, Tamilnadu, India.

Abstract. An organic- inorganic NLO single crystal was grown using conventional and slow cooling method or ATR (Assembled temperature reduction method). Properties of as grown crystal were tried to modify using swift heavy ion (Au^{3+}) of various fluences ranging from 10^{13} ions/cm^2, to 10^{14} ions/ cm^2. Dielectric property of orthorhombic crystal was studied. It was noticed that, as ion fluence increases the dielectric constant of as grown material also increases. Hardness of the undamaged and swift heavy ion irradiated crystals was studied. It was concluded that the ability of resistance of deformation of swift heavy ion implanted crystal increases with increase of ion fluences. Thermal studies were carried out and it was observed that irradiated crystal exhibits increased thermal stability with increase of ion energy and this was owing to decreased degree of crystallinity.

1. Introduction

High energetic ion acts an important role in changing the characterization of materials such as surface, electrical, mechanical thermal etc. It entered deep into the crystals and creates a long and narrow amorphous region along its path [1].Crystals with swift heavy ion (SHI) result to the production of different defects in the crystal system which modifies the physicochemical properties of solids.
The modifications are mostly dependent on the energy of SHI, incident ion mass and temperature in which the irradiation takes place. Ionization and excitation will be caused by heavy ion irradiation and sometimes atoms of target materials will be removed from their lattice [2-4].Swift heavy ion (SHI) interacts with the target materials and it loses its energy in two ways along its trajectory. Swift heavy ion losses its energy with the atomic electron of the target crystal and that is inelastic scattering. Heavy energetic ions loss its energy with the nuclei of the target atoms and it is called elastic scattering [5].
Energy is transferred from heavy energetic ion to crystals and materials. Specific effects are created where SHI transfer its energy to materials [6-7]. Sometimes the refractive index of the crystal changes significantly by SHI [8]. Swift heavy ion capable of breaking molecular bonds and produce damaged create defects in the solids [9]. The efficiency of luminescence increases due to point defects which created during irradiation that act as a colour centre [10-11]. Energetic heavy ion creates

[1]ambroserajkumarphd@gmail.com

amorphization and the degree of crystallinity of the material gets changed when it interacts with target crystal [12-14]. In photonics technology NLO single crystals plays a vital role in fabrication devices. 2-amino-5-nitropyridine (2A5NP) has an attractive molecular structure. 2-amino-5-nitropyridine has nitro and amino group which act as an as an electron acceptor and electron donor respectively which stimulate nonlinear character in the crystal [15-18]. Because of this attractive moleculat structure, 2-amino-5-nitropyridine (2A5NP) is used to construct NLO materials. It is a good choice for second order nonlinear optical (SONLO) applications [19-24]. Number of 2-amino-5-nitropyridine (2A5NP) derivative crystals such as 2-amino-5-nitropyridinium hydrogen oxalate, 2-amino-5-nitropyridinium chloride, 2-amino-5-nitropyridinium nitrate, 2-amino-5-nitropyridinium dihydrogen phosphate and 2-amino-5-nitropyridinium sulfamate were grown using conventional and slow cooling method [25-36]. Structural optical Properties of both undamaged and SHI irradiated 2-amino-5-nitropyridinium sulfamate NLO single were reported [37-38]. Electrical, mechanical, and thermal properties of both pristine and Au^{3+} ion irradiated 2-amino-5-nitropyridinium sulfamate NLO single are reported in this research paper.

2. Experimental

2.1.1 Crystal growth

Conventional method (SET) was used to grow a semiorganic single crystal. Large size crystal of 2-amino-5-nitropyridinium single crystal was grown using temperature reduction method. 2-amino-5-nitropyridine and sulfamic acid were taken in equimolar (1: 1) ratio and were dissolved in distilled water. In order to get homogeneous solution the mixture was stirred for 6 hours. Then the mixture was filtered using Whatman filter paper. To evaporate the water molecules, the filtered mixture was kept at room temperature. A colourless semiorganic single crystal was collected for 90 days [29].

2.1.2 Swift heavy ion ($Au^{3+)}$ irradiation:

A metallic swift heavy ion (Au^{3+}) of varies fluences was used for irradiation with various beam current. Swift heavy ion scanned normal to the face of the crystal uniformly at room temperature. Metallic ion ($Au^{3+)}$ with different beam current 1.55μA (1x 10^{13} ions/cm^2), 1.35 μA (5 x 10^{13} ions/cm^2) and 8 PnA (1x 10^{14} ions/cm^2) were used or irradiation. At room temperature, a metallic ion (Au^{3+}) of 10.8 MeV was used for implantation [38].

3. Results and Discussion

3.1. Microhardness

Hardness is the resistance offered by materials to plastic deformation caused by indentation. Hardness of as grown crystals of pristine and SHI implanted crystals was measured using Vicker's hardness tester The selected smooth surface of 2-amino-5-nitropyridinium sulfamate pristine and irradiated was used to measure the hardness at room temperature. The load was applied on the selected surface ranging from 25 g to 100 g. Four indentations were introduced in the sample of the same surface for each load. Both the diagonal lengths were measured in each indentation marked and the averages of eight diagonals were taken for calculations. Figure 3.1.1 shows the hardness of pristine and heavy ion implanted crystals. It is examined from the graph that both pristine and implanted crystal show linear behavior with applied load. Hardness of the undamaged and implanted crystals increases with applied load. It is also evident from the graph that hardness of the irradiated crystal is significantly larger than the pristine crystals. It is noted that hardness of the irradiated crystal increases with increases of ion fluences. This is due to certain amorphous formed in the irradiated samples. Least square fit method was used to calculated "n" hardening coefficients. According to Onitsch, the crystal is soft if the value of n is greater than 1.6 and the crystal is hard if the value of n is less than 1.6 [39]. Hardening co- efficient was found to be 2.2 for irradiated. So as grown crystals of 2-amino-5-nitropyridinium sulfamate belongs to soft materials. In order to understand stiffness and yield strength of plastic deformation of the pristine and irradiated, stiffness was determined using Wooster's empirical formula [40] and it is given by C_{11}= Hv(7/4). Figure 3.1.2 and 3.1.3 shows stiffness and yield strength of pristine and swift heavy ion irradiated crystals. It is understood from the graph that stiffness and yield strength increase with applied load. It is also elucidated that stiffness and yield strength of irradiated crystal is significantly larger than the pristine sample. It is noticed from the graph that stiffness and yield strength show linear behaviour with applied load. Stiffness and yield strength increase with increase of ion fluence from 10^{13}ions/cm^2 to 10^{14}ions/cm^2. Increase of binding force between ions and mechanical strength of Au^{3+} ion with three different fluences is due to defects created in the crystal which makes the crystalline material into amorphous nature.

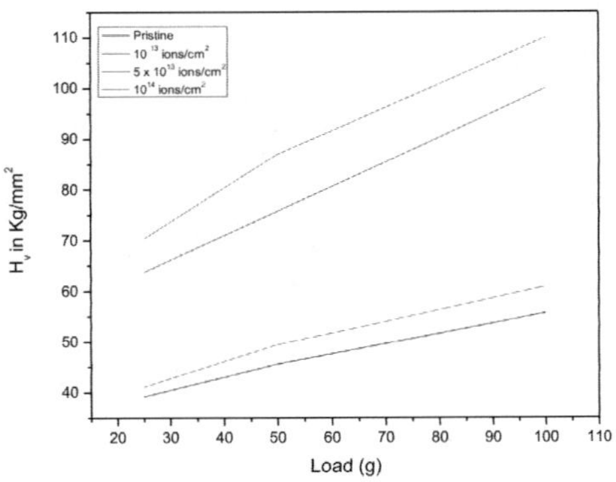

Fig.3.1.1 Hardness graph of pristine and Au^{3+} ion irradiated 2- amino-5-nitropyridinium sulfamate (2A5NPS) NLO single crystal.

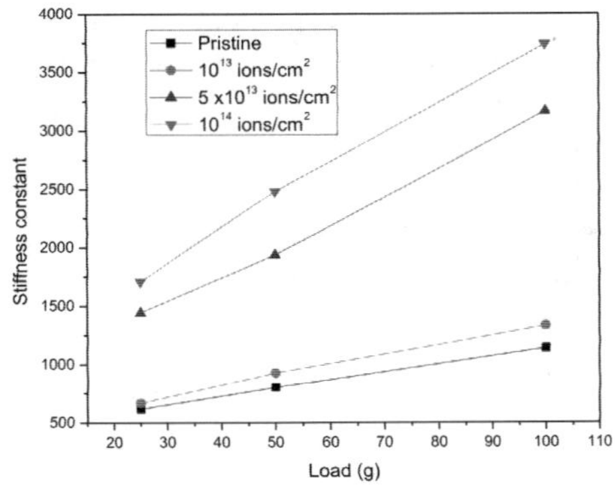

Fig.3.1.2 Stiffness graph of pristine and Au^{3+} ion irradiated 2-amino-5-nitropyridinium sulfamate (2A5NPS) NLO single crystal.

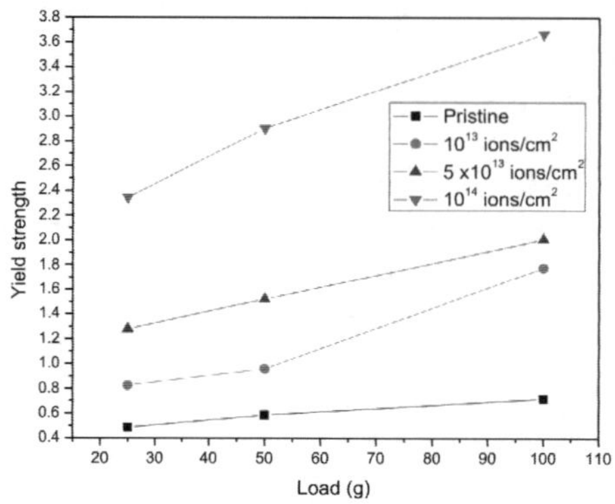

Fig.3.1.3 Yield strength of pristine and Au^{3+} ion irradiated 2-amino-5-nitropyridinium sulfamate (2A5NPS) NLO single crystal.

3.2. Dielectric study

From graph (Fig. 3.2.1), it is observed that dielectric constant decreases up to a certain frequency range and remains constant. Dielectric constant is high at lower frequency due to ionic, electric, orientation and space charge polarization. The dielectric constant value is high for a swift heavy ion irradiated crystals than that of pristine sample. It also is noted that dielectric constant increases as ion fluence increase from 10^{13} ions/cm^2 to 10^{14} ions/cm^2. The increased dielectric constant in swift heavy ion irradiated crystals is the formation of defect along the ion trajectory. The metallic ion which is implanted in the orthorhombic crystals loses energy by elastic and inelastic collision. The crystal was damaged and the lattice was disordered by the swift heavy ion (SHI) which lost its energy in the form of collision. Irradiated ion which lose energy in the crystal in the form of collision which damage the crystals and disordering the lattice. The damaged zones are separated from each other as the ion produces a region of amorphous surrounded by regions containing defects. Dielectric constant increases as increase of ion fluences which produce lattice disorderness. Due to lattice disorderness more ions are activated which brings more interaction between ions results increase the capacitance and hence dielectric constant increases.

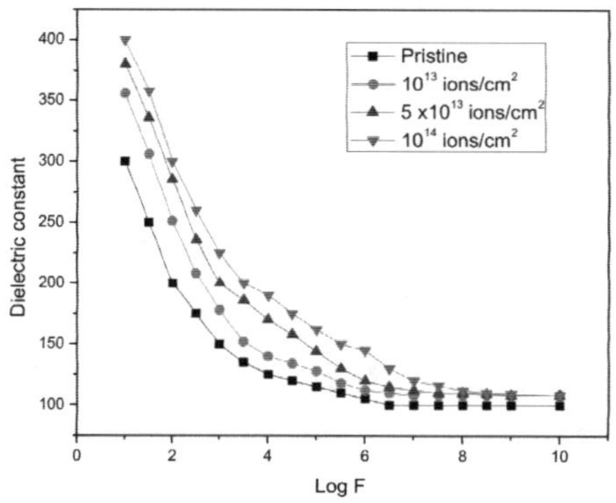

Fig.3.2.1 Dielectric constant of pristine and Au^{3+} ion irradiated 2-amino-5-nitropyridinium sulfamate (2A5NPS) NLO single crystal.

3.3 Thermal Analysis

TG and DTA analysis give information about physical phenomena, such as phase transition, oxidation, reduction and chemical phenomena including thermal decomposition. TG and DTA was studied for undamaged and SHI irradiated 2-amino-5-nitropyridinium crystals. It is observed from the graph (Fig. 3.3 a, b, c) that samples decompose in three different stages. A pristine sample was started to decompose at 195°C and an endothermic peak is absorbed in DTA cure at 195°C which confirm the melting point of undamaged sample. Au^{3+} ion irradiated single crystal with different fluences (1 x10^{13} ions/cm^2, 5 x 10^{13} ions/ cm^2 , and 1 x10^{14} ions/ cm^2) were decomposed with weight loss was little more than the pristine at 196^0 C, 196.5^0 C and 198^0 C respectively. There was no endothermic peak observed before decomposition which is the melting point of crystal system. Weight loss of pristine and irradiated sample increases with increase of temperature after melting of the crystal. Pristine and irradiated crystals decompose almost completely on or before 300° C. Thermal stability increased due to amphorphus surface created after SHI irradiation. It is also noted that thermal stability increases with increase of ion fluences. This is due to increase of defects which are caused by increase of ion fluences.

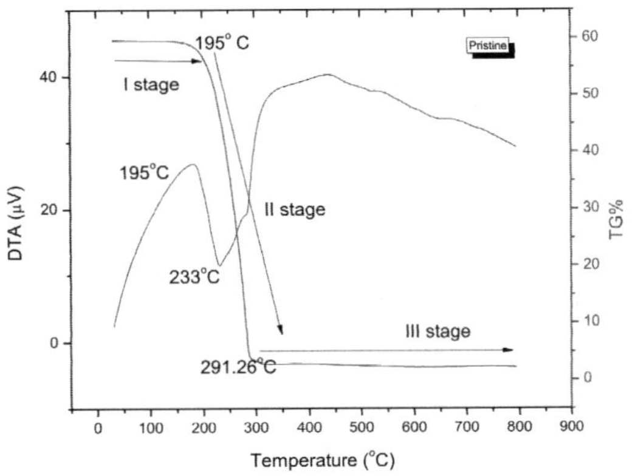

Fig.3.3 a) Thermal analysis of 2-amino-5-nitropyridinium sulfamate (2A5NPS) NLO single crystal.

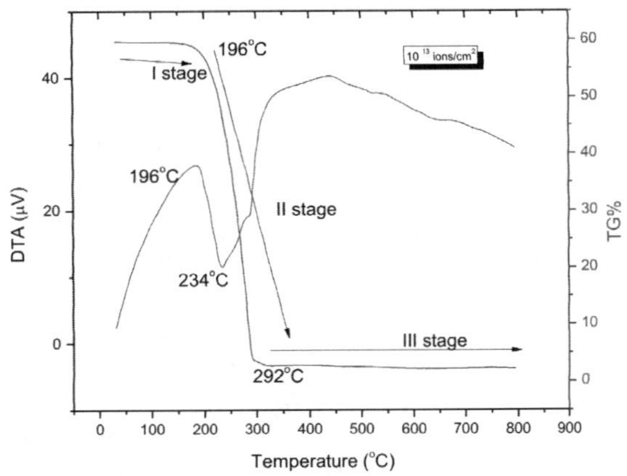

Fig.3.3 b) Thermal analysis of Au^{3+} ion irradiated (10^{13}ions/cm^2) 2-amino-5-nitropyridinium sulfamate (2A5NPS) NLO single crystal.

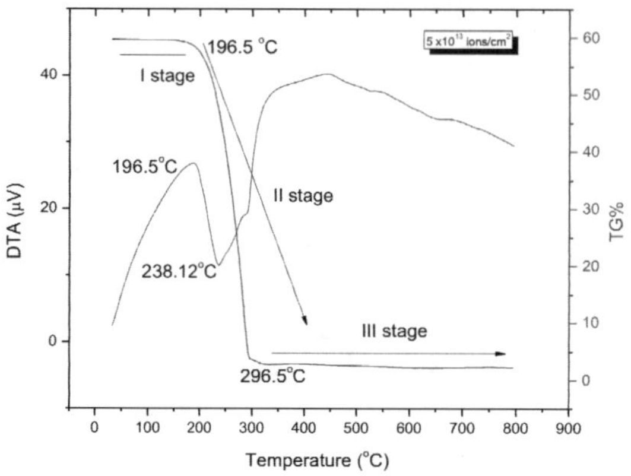

Fig.3.3 c) Thermal analysis of Au^{3+} ion irradiated (5x 10^{13}ions/cm^2) 2-amino-5-nitropyridinium sulfamate (2A5NPS) NLO single crystal.

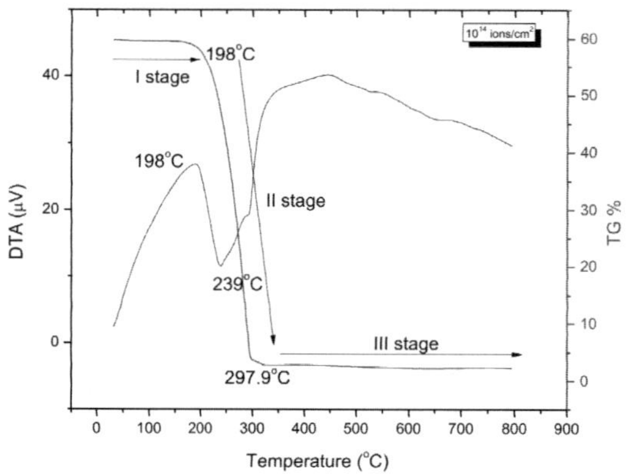

Fig.3.3 d) Thermal analysis of Au^{3+} ion irradiated (10^{14} ions/cm^2) 2-amino-5-nitropyridinium sulfamate (2A5NPS) NLO single crystal.

4. Conclusion

Orthorhombic unit of 2-amino-5-nitropyridinium sulfamate (2A5NPS) was grown using conventional evaporation method and Assembled Temperature Reduction (ATR) method. It was observed from the dilectric studies that Dielectric constant increases as increase of ion fluences which produce lattice disorderness. Due to lattice disorderness more ions are activated which brings more interaction between ions results increase the capacitance and hence dielectric constant increases. Microhardness of both pristine and irradiated was measured using Vicker's hardness test. It was concluded from hardness that hardness of the irradiated crystal increased with increases of ion fluences. This was due to certain amorphous formed in the irradiated samples. Stiffness and yield strength increased with increases of ion fluence from 10^{13}ions/cm^2 to 10^{14}ions/cm^2. Increase of binding force between ions and mechanical strength of Au^{3+} ion with three different fluences was due to defects created in the crystal which made the crystalline material into amorphous nature. Thermal studies were carried out and thermal stability increased due to amphorphus surface created after SHI irradiation. It is also noted that thermal stability increases with increase of ion fluences. This is due to increase of defects which are caused by increase of ion fluences.

5. References

[1]. Kanjilal. D, 2001, ‘Swift heavy ion-induced modification and track formation in materials’ Curr.sci, Vol. 80, PP.1560.
[2]. Kucheyev. S. O, Williams J.S, Zou. J, Jagadish. C & Li. G, 2001, ‘The effects of ion mass, energy, dose, flux and irradiation temperature on implantation disorder in GaN’, Nucl. Instrum. Methods B, Vol.178, PP. 209-213.
[3]. Kiritani. M, 1998, ‘The need for improved temperature control during reactor irradiation’ J. Nucl.Mater, Vol.160, PP. 135.
[4]. Seager. C.H, Fleming. R.M, Lang. D.V, Cooper. P.J, Bielejec. E, & Campbell. J.M, 2007, ‘Effects of defect clustering in neutron irradiated silicon’, Physica B: Condensed Matter, Vol. 401–402, PP. 491–494.
[5]. Zhang. X & Xue. D, 2007, ‘Bond Energy Prediction of Curie Temperature of Lithium Niobate Crystals’, Journal of Physical Chemistry B, Vol.111, PP. 2587.
[6]. Vorobyova. I.V & Kopnicxky. J, 2005, MeV-atomic-ion-induced tracks on the surface of CaF2 single crystal Nuc. Instr.and Meth.in Phy. Res.Sec.B: Beam Interactions with Materials and Atoms, Vol. 237, PP. 593-601
[7]. Lesuer. O & Dunlop. A,1993, ‘Damage creation via electronic excitations in metallic targets part I: Experimental results’, Radi. eff. defect in solids, Vol. 126, PP. 123-128.
[8]. Sreeramana Aithal. P, Nagaraja. H.S, Mohan Rao. P, Avasthi. D.K, & Asati Sarma, 1997 ‘Effect of high energy ion irradiation on electrical and optical properties of para-hydroxy acetophenone’, J. Appl. Phys. 81, PP 7526-7528.
[9]. Nagabhushana. H, Prashantha. S.C, Nagabhushana. B.M, Lakshminarasappa. B.N, Fouran Singh, & Chakradhar. R.P.S,2008a 'Ion beam-induced luminescence and photoluminescence of 100 MeV Si8+ ion irradiated kyanite single crystals', Solid Status Commun. 147, 9-10, PP. 377-380.
[10]. Deepthy, K.S.R. Rao, H.L. Bhat, Ravikumar, & Asokan, K, 2001, ‘Gray track formation in KTiOPO4 by swift ion irradiation’, J. Appl. Phy, Vol.89, PP.6560-6562.
[11]. Kanagasekaran. T, Mythili. P, Bhagavannarayana. G, Kanjilal, & D, Gopalakrishnan. R, 2009, ‘Investigations of structural, dielectric and optical properties on silicon ion irradiated glycine monophosphate single crystals’, Nucl. Instr. Methods Phys. Res. B 267, PP 2495-2502.
[12]. Li Fucin, Luping, Himaru Manbuies, Kurt E. Sicafus, & Philos, 2000, ‘Effect of heavy ion irradiation on near-surface microstructure in single crystals of rutile TiO_2’, Mag.B, Vol. 80, PP.1947-1954.
[13]. Sun. Y., Zhang. C, Zhu. Z, Wang. Z, Jin. Y, Liu. J & Wang. Y, 2006, ‘Evolution of radiation-induced carbon–oxygen-related defects in silicon upon annealing: LVM studies’,Nuc. Inst. and Meth. Phy. Res. Sec.B: Beam Interactions with Materials and Atoms, Vol. 245, PP. 210-213.
[14]. Olivares.J, García-Navarro. A, García. G, Agulló-López. F, Agulló-Rueda. F, García-Cabañes. A & Carrascosa. M, 2007, ' Nonlinear optical waveguides generated in lithium niobate by swift-ion irradiation at ultralow fluences ', Optics letters,Vol.32, 17, PP. 2587-2589.
[15]. Yogam. F, Vetha Potheher. I, Cyrac Peter. A, Tamilselvan. S, Leo Rajesh. A, Vimalan. M & Sagayaraj. P, 2011, ‘Growth and characterization of novel semiorganic nonlinear optical crystals of L-phenylalanine hydrochloride (LPHCl) ’, Advances in Applied Science Research, Vol. 2, 1, PP. 261-268.
[16]. Prasad. P.N. and William. D.J, 1991, Wiley, Newyork,
17. Masse. R, Bagien- Beucher. M, Pecaut. J, Levy. J.P, Zyss. J, 1993. J. Nonlinear Opt.Vol.5, PP.413–423.
[18]. Pecaut. J, & Masse. R, 1993, 'Structure of bis(2-amino-5-nitropyridinium) dichromate as a step towards the design of efficient organic-inorganic non-linear optical crystals', Acta Crystallogr. B Vol.49 ,PP. 277–282.
[19]. Masse. R, & Zyss. J,1991, 'A new approach in the design of polar crystals for quadratic nonlinear optics exemplified by the synthesis and crystal structure of 2-amino-5-nitropyridinium dihydrogen monophosphate (2A5NPDP)', Mol. Eng.Vol.1, 2, PP. 141–152.

[20]. Horiuchi. N, Lefaucheux. F, Robert. M.C, Jose. D, Khodja. S, & Zyss. J, 1995,' Gel growth of 2-amino-5nitropyridinium dihydrogen phosphate organomineral crystals: X-ray and nonlinear optics characterization', Journal of crystal growth 147, PP. 361-368.
[21]. Dhanuskodi. S, Pricilla Jeyakumari. A, & Manivannan. S, 'Semiorganic NLO Material for Short Wavelength Generation: 2-amino-5-nitro Pyridinium Bromide', Journal of crystal growth, Vol.282, PP.72-78, 2005.
[22]. Anandha Babu. G, Perumal Ramasamy. R, Ramasamy. P & Krishna Kumar. V, 2009, ' Synthesis, Crystal Growth, and Characterization of an Organic Nonlinear Optical Donor-π-Acceptor Single Crystal: 2-amino-5-nitropyridinium-toluenesulfonate', Cryst. Growth Des, Vol.9, 7, PP. 3333–3337.
[23]. Anna Puig-Molina, Angel Alvarez-Larena, Juan. F. Piniella, Sean. T. Howard & François Baert, 1998, 'The Electron Distribution in the Nonlinear Optical Material 2-Amino-5-Nitropyridinium Dihydrogen Phosphate, Structural Chemistry, Vol. 9, No. 6, PP. 395–402.
[24]. Arivazhagana. M & Subhasinib. V.P, 2012, 'Quantum chemical studies on structure of 2-amino-5-nitropyrimidine', Spectrochimica Acta Part A 91, PP. 402– 410.
[25]. Rodrigues. V. H, Costa. M. M. R. R, Dekola. T & de Matos Gomes. E, 2009, '2-Amino-5-nitropyridinium tetraoxidorhenate(VII) monohydrate', Acta Cryst. E65, 2009.
[26]. Lin Chen, Yang Cao, Tiantian Fang, Wei Yue, Jing Wang, Xin Zhang, & Fengchun Yang, 2013, 'Synthesis and characterization of degradable polyimides from p-phenylenedioxybis(5-amino-2-pyridine)', Polymer Degradation and Stability, Vol. 98, PP. 839- 843.
[27]. Sivasubramani. V, Senthil Pandian. M, & Ramasamy. P, 2015, 'Studies on 2- amino 5-nitropyridinium nitrate (2A5NPN) a semi- organic third order nonlinear optical single crystal' AIP conference, 150, PP 765-771, 2015
[28]. Ambrose Rajkumar M, Stanly John Xavier. X, Anbarasu. S, & Prem Anand. D, 2014(a),' Photoconductivity, Microhardness and dielectric studies of 2-amino-5- nitropyridinium hydrogen oxalate (2A5NPHO) single crystal', International Journal of Emerging Trends in Science and Technology Vol.1, issue.10, PP. 1579- 1585.
[29]. Ambrose Rajkumar. M, Stanly John Xavier. X, Anbarasu.S, & Prem Anand. D, 2016(a), 'Growth and characterization studies of an efficient semiorganic NLO single crystal: 2-amino-5-nitropyridinium sulfamate (2A5NPS) by assembled temperature reduction (ATR) method', Optical Materials, Vol. 55, PP.153–159.
30. Ambrose Rajkumar. M, Stanly John Xavier. X, Anbarasu.S, & Prem Anand. D, 2014, '2-amino-5-nitropyridinium hydrogen oxalate', Acta Cryst. E70,PP. o473–o474,
[31]. Ambrose Rajkumar. M, Stanly John Xavier. X, Anbarasu.S, & Prem Anand. D, 2014(b), 'Microhardness, dielectric and photoconductivity, studies of 2-amino-5- nitropyridinium dihydrogen phosphate (2A5NPDP) nlo single crystal', International Journal of Emerging Trends in Science and Technology Vol.1, issue.10, PP.1610-1615.
[32]. Ambrose Rajkumar. M, Stanly John Xavier. X, Anbarasu.S, & Prem Anand. D, 2014(c), 'A Convenient Route to Synthesize and Grow 2-amino-5- nitropyridinium nitrate crystals for laser generation 'Sciencia Acta Xaveriana, Vol.4, 2, PP.73-78.
[33]. Ambrose Rajkumar. M, Stanly John Xavier. X, Anbarasu.S, & Prem Anand. D, 2014(d), 'Microhardness, Dielectric and photoconductivity, studies of 2-amino-5- nitropyridinium nitrate (2A5NPN) NLO single crystal', Res. J. Physical Sci, Vol. 2(1), PP.1-4.
[34]. Ambrose Rajkumar. M, Stanly John Xavier. X, Anbarasu.S, & Prem Anand. D, 2015, 'Crystal structure of 2-amino-5-nitropyridinium sulfamate,'Acta Cryst. E71.
[35]. Ambrose Rajkumar. M, Stanly John Xavier. X, Anbarasu.S, & Prem Anand, D,2016, ' Growth and characterization studies of an efficient semiorganic NLO single crystal: 2-amino-5-nitropyridinium dihydrogen phosphate (2A5NPDP) by Sankaranarayanan- Ramasamy method', Optik Vol.127, PP. 2187-2192.
[36]. Ambrose Rajkumar. M, Stanly John Xavier. X, Anbarasu.S, & Prem Anand,D, 2017, 'Crystal growth and characterizations of an efficient semiorganic nonlinear optical (NLO) single crystal: 2-amino-5-nitropyridinium chloride (2A5NPCl) by assembled temperature reduction apparatus (ATR) method', Materials research innovations.

IOP Conf. Series: Materials Science and Engineering 1124 (2021) 012005 doi:10.1088/1757-899X/1124/1/012005

[37]. M. Ambrose Rajkumar, G. Prathap, P. Ajith, S. Anbarasu, Prem Anand Devarajan , 2021'Swift heavy ion (Au3+) irradiated 2-amino-5-nitropyridinium chloride (2A5NPCl) nonlinear optical (NLO) single crystal, IJRAR January 2021, Volume 8, Issue 1
[38]. M. Ambrose Rajkumar, A. Ajeeth, S. Anbarasu, and Prem Anand Devarajan, Irradiation effect of Au^{3+} on 2-amino-5- nitropyridinium sulfamate (2A5NPS) NLO single crystal, AIP Conference Proceedings 2270, 100002 (2020)
[39]. Onitsch. E.M., 1947, Mikroskopie Vol.2, 131-151.
[40]. Wooster. W.A, 1953, 'Physical properties and atomic arrangements in crystals', Reports on Progress in Physics, Vol. 16, PP. 62-82.

AUTHOR INDEX

Institute of Physics
Dirac House, Temple Back
Bristol BS1 6BE UK

ISSN: 1757-8981
ISBN 978-1-7138-3407-6

CURRAN ASSOCIATES INC.
proceedings.com